Apoptosis

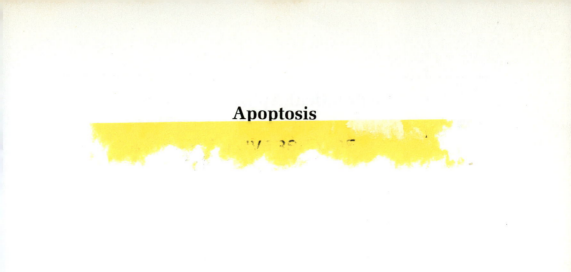

The Practical Approach Series

SERIES EDITOR

B. D. HAMES
Department of Biochemistry and Molecular Biology
University of Leeds, Leeds LS2 9JT, UK

See also the Practical Approach web site at **http://www.oup.co.uk/PAS**

★ **indicates new and forthcoming titles**

Affinity Chromatography
Affinity Separations
Anaerobic Microbiology
Animal Cell Culture
   (2nd edition)
Animal Virus Pathogenesis
Antibodies I and II
Antibody Engineering
Antisense Technology
★ Apoptosis
Applied Microbial Physiology
Basic Cell Culture
Behavioural Neuroscience
Bioenergetics
Biological Data Analysis
Biomechanics – Materials
Biomechanics – Structures and
   Systems
Biosensors
★ Caenorhabditis Elegans
Carbohydrate Analysis
   (2nd edition)
Cell-Cell Interactions
The Cell Cycle

Cell Growth and Apoptosis
★ Cell Growth, Differentiation
   and Senescence
★ Cell Separation
Cellular Calcium
Cellular Interactions in
   Development
Cellular Neurobiology
Chromatin
★ Chromosome Structural
   Analysis
Clinical Immunology
Complement
★ Crystallization of Nucleic
   Acids and Proteins
   (2nd edition)
Cytokines (2nd edition)
The Cytoskeleton
Diagnostic Molecular
   Pathology I and II
DNA and Protein Sequence
   Analysis
DNA Cloning 1: Core
   Techniques (2nd edition)
DNA Cloning 2: Expression
   Systems (2nd edition)

Apoptosis

A Practical Approach

Edited by

GEORGE P. STUDZINSKI

Department of Pathology and Laboratory Medicine
UMD—New Jersey Medical School
Newark, N.J., USA

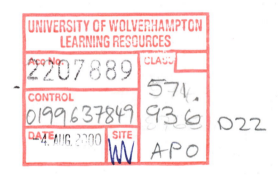

OXFORD
UNIVERSITY PRESS

OXFORD
UNIVERSITY PRESS

Great Clarendon Street, Oxford OX2 6DP

Oxford University Press is a department of the University of Oxford
and furthers the University's aim of excellence in research, scholarship,
and education by publishing worldwide in

Oxford New York

Athens Auckland Bangkok Bogotá Buenos Aires Calcutta
Cape Town Chennai Dar es Salaam Delhi Florence Hong Kong Istanbul
Karachi Kuala Lumpur Madrid Melbourne Mexico City Mumbai
Nairobi Paris São Paulo Singapore Taipei Tokyo Toronto Warsaw

and associated companies in Berlin Ibadan

Oxford is a registered trade mark of Oxford University Press

Published in the United States
by Oxford University Press Inc., New York

© Oxford University Press, 1999

Users of books in the Practical Approach Series are advised that prudent
laboratory safety procedures should be followed at all times. Oxford
University Press makes no representation, express or implied, in respect of
the accuracy of the material set forth in books in this series and cannot
accept any legal responsibility or liability for any errors or omissions
that may be made.

A catalogue record for this book is available from the British Library

Library of Congress Cataloging in Publication Data
(Data available)

ISBN 0-19-963784-9 (Hbk)
       0-19-963785-7 (Pbk)

Typeset by Footnote Graphics,
Warminster, Wilts
Printed in Great Britain by Information Press, Ltd,
Eynsham, Oxon.

Preface

Apoptosis, a concept derived from pathological observations dating back to the availability of the microscope, has been recently afforded treatment reminiscent of the ancient apocalyptic writings. Revelations are reported daily of new, and so it is claimed, profound insights into cellular survival mechanisms, and their principal default pathway, apoptosis. The preoccupation of many scientists with this cellular programme, or programmes, appears to be driven by several currents. There is a bewildering array of commercial reagents and accompanying literature reporting to provide, often quick and easy, means of discovering important secrets of nature. There is also the fascination of the scientists with the beauty of an almost endless cascades of protein–protein interactions that lead to an irrevocable end-point—cell death. And of course, there is the legitimate expectation that important components of therapy for cancer and immune diseases will be based on our understanding of the precise mechanisms of these apoptotic cascades.

This volume presents the techniques essential for contemporary research on diverse aspects of apoptosis. In addition to the basic methodology for recognition of the apoptotic phenotype and its characteristic DNA fragmentation, the text contains a wide variety of procedures used to investigate the mechanistic aspects of the programmes for survival or death of mammalian cells. A team of scientists who are among the leaders in apoptosis research has provided numerous protocols which describe in detail how to perform these procedures and discusses them from the individual points of view of each contributor. The protocols most frequently used in current investigations of apopotosis research are presented with variations that have been found particularly useful for a particular application, thus allowing the reader to benefit from the experience of laboratories which focus on different aspects of apoptosis research.

Attention is also directed to the choice of the procedures, to pitfalls in their execution, and to critical interpretation of the results. It is believed that the nuances of technical approaches discussed here will be helpful to the experienced as well as the beginning investigators. The credit for this must go to the team of authors and the OUP staff.

New Jersey   G.P.S.
1999

Contents

5. Cell-mediated cytotoxicity and cell death receptors 81

6. Sphingolipids as messengers of cell death 105

7. Cytochemical detection of cytoskeletal and nucleoskeletal changes during apoptosis   125

Manon van Engeland, Bert Schutte, Anton H. N. Hopman, Frans C. S. Ramaekers, and Chris P. M. Reutelingsperger

8. Metabolic alterations associated with apoptosis   141

Ana P. Costa-Pereira and Thomas G. Cotter

Contents

9. Methods of measuring Bcl-2 family proteins and their functions 157

John C. Reed, Zhihua Xie, Shinichi Kitada, Juan M. Zapata,
Qunli Xu, Sharon Schendel, Maryla Krajewska, and
Stanislaw Krajewski

Contents

10. Methods for detecting proteolysis during apoptosis in intact cells 215

April L. Blajeski and Scott H. Kaufmann

Contributors

ALBERTO ANEL
Department Bioquimica y Biologia Molecular y Celular, Faculty of Sciences, University of Zaragoza, Zaragoza, 50009, Spain

ELZBIETA BEDNER
Brander Cancer Research Institute, 19 Bradhurst Avenue, Hawthorne, NY., 10532, USA

APRIL L. BLAJESKI
Department of Pharmacology, Mayo Graduate School, Rochester, MN 55905, USA

ANA P. COSTA-PEREIRA
Tumour Biology Laboratory, Department of Biochemistry, University College, Cork, Ireland

THOMAS G. COTTER
Tumour Biology Laboratory, Department of Biochemistry, University College, Cork, Ireland

ZBIGNIEW DARZYNKIEWICZ
Brander Cancer Research Institute, 19 Bradhurst Avenue, Hawthorne, NY., 10532, USA

YUSUF A. HANNUN
114 Doughty St, Rm. 603, Strom Thurmond Bldg, Charleston, SC 29425, USA

ANTON H. N. HOPMAN
Department of Molecular Cell Biology and Genetics, University of Maastricht, PO Box 616, 6200 MD Maastricht, The Netherlands

GARY JENKINS
114 Doughty St, Rm. 603, Strom Thurmond Bldg, Charleston, SC 29425, USA

SCOTT H. KAUFMANN
Guggenheim 1342C, Mayo Clinic, 200 First St., S.W., Rochester, MN 55905, USA

SHINICHI KITADA
The Burnham Institute, 10901 North Torrey Pines Road, La Jolla, CA, 93037, USA

MARYLA KRAJEWSKA
The Burnham Institute, 10901 North Torrey Pines Road, La Jolla, CA, 93037, USA

Contributors

STANISLAW KRAJEWSKI
The Burnham Institute, 10901 North Torrey Pines Road, La Jolla, CA, 93037, USA

XUN LI
Brander Cancer Research Institute, 19 Bradhurst Avenue, Hawthorne, NY., 10532, USA

JAVIER NAVAL
Department Bioquimica y Biologia Molecular y Celular, Faculty of Sciences, University of Zaragoza, Zaragoza, 50009, Spain

YVES POMMIER
Laboratory of Molecular Pharmacology, Division of Basic Sciences, National Cancer Institute, National Institute of Health, Bethesda, Maryland, USA

CHRISTOPHER S. POTTEN
CRC Epithelial Biology Laboratory, Section of Cell and Tumour Biology, Paterson Institute for Cancer Research, Wilmslow Road, Withington, Manchester M20 4BX, UK

FRANS C. S. RAMAEKERS
Department of Molecular Cell Biology and Genetics, University of Maastricht, PO Box 616, 6200 MD Maastricht, The Netherlands

JOHN C. REED
The Burnham Institute, 10901 North Torrey Pines Road, La Jolla, CA, 93037, USA

CHRIS P. M. REUTELINGSPERGER
Department of Biochemistry, University of Maastricht, PO Box 616, 6200 MD Maastricht, The Netherlands

SHARON SCHENDEL
The Burnham Institute, 10901 North Torrey Pines Road, La Jolla, CA, 93037, USA

BERT SCHUTTE
Department of Molecular Cell Biology and Genetics, University of Maastricht, PO Box 616, 6200 MD Maastricht, The Netherlands

RONG-GUANG SHAO
Department of Oncology, Institute of Medicinal Biotechnology, Chinese Academy of Medical Sciences, 1 Tiantan xili, Beijing 100050, P. R. China

GEORGE P. STUDZINSKI
Department of Pathology and Laboratory Medicine, UMDNJ-New Jersey Medical School, 185 S. Orange Avenue, University Heights, Newark, NJ 07103, USA

Contributors

MANON VAN ENGELAND
Department of Molecular Cell Biology and Genetics, University of Maastricht, PO Box 616, 6200 MD Maastricht, The Netherlands

JAMES W. WILSON
CRC Epithelial Biology Laboratory, Section of Cell and Tumour Biology, Paterson Institute for Cancer Research, Wilmslow Road, Withington, Manchester M20 4BX, UK

ZHIHUA XIE
The Burnham Institute, 10901 North Torrey Pines Road, La Jolla, CA, 93037, USA

QUNLI XU
Oncology Disease Group, Hoechst & Marion Roussel, Inc., Route 202–206, Bridgewater, NJ 08807, USA

AKIRA YOSHIDA
First Department of Internal Medicine, Fukui Medical School, Shimoaizuki 23, Matsuoka-cho, Fukui, 910–1193, Japan

Abbreviations

^{125}IUdR	^{125}I-deoxyuridine
2-VP	2-vinylpyridine
ABC	avidin–biotin complex
Ac-DEVD-CHO	acetyl-Asp-Glu-Val-Asp-aldehyde
Ac-YVAD-cmk	acetyl-Tyr-Val-Ala-Asp-chloromethylketone
AEC	3-amino-9-ethylcarbasole
AFC	7-amino-4-trifluoromethylcoumarin
AI	apoptotic index
AICD	activation-induced cell death
AID	activation-induced death
ALS	alkali-labile sites
AMC	7-amino-4-methylcoumarin
APAF	apoptotic protease activating factor
APES	3-aminopropyl triethoxy silane
B-CLL	B cell chronic lymphocytic leukaemia
BCA	bicinchoninic acid
BCIP	bromochloroindolyl phosphate
BLT	N-benzyloxycarbonyl-L-lysine thiobenzyl ester
Boc-D-fmk	t-butyloxycarbonyl-Asp-fluoromethylketone
βOG	β-octyl glucoside
bp	base pair
BSA	bovine serum albumin
BSO	DL-buthionine-S,R-sulfoximine
CAD	caspase-activated DNase
CAPK	ceramide-activated protein kinase
CAPP	ceramide-activated protein phosphatase
CARD	caspase recruitment domain
CD	cluster designation
CHEF	clamped homogeneous electric field
CoA	co-enzyme A
CPT	comptothecin
CTL	cytotoxic T lymphocyte
Cu/ZnSOD	copper/zinc superoxide dismutase
D-FPR-cmk	D-Phe-Pro-Arg-clhoromethylketone
DAB	3′3′-diaminobenzidine
DAG	diacylglycerol
DAPI	4′,6-diamidino-2-phenylindole
DCFH/DA	2′,7′-dichlorofluorescein diacetate
DCI	3,4-dichloroisocoumarin

DED	death effector domains
DEP	diethyl pyrocarbonate
DFF	DNA fragmentation factor
DGK	diacylglycerol kinase
DHE	dihydroethidium
DMF	dimethylformamide
DMG	dimethyl glutaric acid
DMSO	dimethyl sulfoxide
DOPG	dioleoylphosphatidlylglycerol
DPC	DNA–protein cross-links
DSB	double-stranded breaks
DTNB	5,5′-dithio-*bis*(2-nitrobenzoic acid)
DTT	dithiothreitol
dUTP	deoxyuridine triphosphate
$\Delta\psi_m$	mitochondrial transmembrane potential
E:T	effector to target ratio
ECL	enhanced chemiluminescence
EDTA	ethylenediaminotetra-acetic acid
EGTA	ethylene glycol tetra-acetic acid
ERK	extracellular signal regulated kinases
FACS	fluorescence-activated cell sorting
FADD	Fas-associated death domain
FAK	focal adhesion kinase
FCS	fetal calf serum
FITC	fluoroscein isothiocyanate
GFP	green fluorescent protein
GSH	glutathione
GSSG	glutathione disulfide
GST	glutathione S-transferase
Gu HCl	guanidine hydrochloride
H&E	haematoxylin and eosin
H_2O_2	hydrogen peroxide
HBSS	Hank's balanced salt solution
HIV	human immunodeficiency virus
HMW	high molecular weight
HO·	hydroxyl radical
HPLC	high performance liquid chromatography
HRPase	horseraddish peroxidase
ICAD	inhibitor of CAD
ICE	interleukin-1β-converting enzyme
IGA	7-(phenyl-ureido)-4-chloro-3-(2-isothioureidoethoxy)-isocoumarin
IPTG	isopropylthio-β-D-galactoside
ISC	interstrand DNA cross-links

ISEL	*in situ* end labelling
ISNT	*in situ* nick translation
JC-1	5,5′,6,6′-tetrachloro-1,1′,3,3′-tetraethyl-benzimidazolylcarbocyanine iodide
kb	kilobase pair
LB	Luria-Bertani broth
LSC	laser-scanning cytometry
LUV	large unilamellar vesicles
MAD	multiple antigen detection procedures
mBCl	monochlorobimane
MHC	major histocompatibility complex
MI	mitotic index
MLC	mixed lymphocyte culture
MnSOD	manganese superoxide dismutase
mPBS	modified PBS
MTT	3-(4,5-dimethylthiazol-2-yl)-2,5-diphenyl-tetrazolium bromide
mt DNA	mitochondrial DNA
NBT	nitroblue tetrazolium
NEM	*N*-ethylmaleimide
NK	natural killer
nu DNA	nuclear DNA
o-PA	*ortho*-phthaldialdehyde
$O_2^{\cdot-}$	superoxide anion
OCT	optimal cutting temperature
OTC	outside tissue compound
PAGE	polyacrylaminde gel electrophoresis
PAP	peroxidase anti-peroxidase complex
PASB	protein-associated strand breaks
PBS	phosphate-buffered saline
PHA	phytohaemagglutinin
PI	propidium iodide
PKC	protein kinase C
PMSF	α-phenylmethylsulfonyl fluoride
pNA	*p*-nitroaniline
PPDA	*p*-phenylenediamine
PS	phosphatidylserine
PT	permeability transition
PVC	polyvinal chloride
PVDF	polyvinylidene fluoride
ROI	reactive oxygen intermediates
RT-PCR	reverse transcriptase-polymerase chain reaction
SDS	sodium dodecyl sulfate
SE	standard error

SG	designation of a commercial chromogenic substrate
SM	sphingomyelin
SMase	sphingomyelinase
SSA	5-sulfosalicylic acid
SSB	single-strand breaks
SSC	sodium chloride (0.15M), trisodium citrate (0.015 M), buffer (pH 7.0)
SUV	small unilamellar vesicles
TAE	Tris–acetate buffer
TBE	Tris–borate buffer
TCR	T cell antigen receptor
TdT	terminal deoxynucleotidyl transferase
TEM	transmission electron microscopy
TLC	thin layer chromatography
TLCK	N-p-tosyl-L-lysine chloromethyl ketone
TMB	3,3′, 5, 5′-tetramethyl benzidene
TNB	5-thio-2-nitrobenzoic acid
TNF	tumour necrosis factor
TPCK	N-α-tosyl-L-phenylalanine chloromethyl ketone
TUNEL	terminal deoxynucleotidyl transferase (TdT)-mediated dUTP nick end-labelling
YVK(bio)D-aomk	N-(acetyltyrosinylvalinyl-N^ε-biotinyllysyl) aspartic acid [(2,6-dimethylbenzoyl)oxy]methyl ketone
Z-AAD-cmk	benzyloxycarbonyl-Ala-Ala-Asp-clhoromethylketone
Z-EK(bio)D-aomk	N-(N^α-benzyloxycarbonylglutamyl-N^ε-biotinyllysyl)aspartic acid [(2,6-dimethylbenzoyl)oxy]methyl ketone
Z-VAD-fmk	benzyloxycarbonyl-Val-Ala-Asp-fluoromethylketone
Z-VDVAD-fmk	benzyloxycarbonyl-Val-Asp-Val-Ala-Asp-fluoromethylketone

Overview of apoptosis

GEORGE P. STUDZINSKI

1. General introduction and overview of contents

When cells receive mixed signals for growth they usually die. For instance, when the developmental programme requires cell division but external growth signals are lacking, or when a growth-related gene such as c-*myc* is highly expressed but the cellular environment has insufficient nutrient content, or a toxic xenobiotic is present, the cell dies by a process termed apoptosis. Although there are differences in the phenomena observed during the apoptotic sequence of events, depending on the cell type, and agent or circumstance which initiates the cell's demise, there are morphological and biochemical similarities which suggest that these are variants of the same biological process, designed to control the size of cell populations.

It is important to distinguish apoptosis from the other major form of cell death, necrosis. First, at the tissue level, apoptosis produces little or no inflammation, since shrunken portions of the cell are engulfed by the neighbouring cells, especially macrophages, rather than being released into the extracellular fluid. In contrast, in necrosis, cellular contents are released into the extracellular fluid, and thus have an irritant effect on the nearby cells, causing inflammation. Secondly, there is the expectation that elucidation of the steps of the cellular mechanisms that lead to apoptosis may allow this form of cell death to be induced more effectively by cancer therapeutic agents. Thirdly, the apoptotic mechanism of cell death is fundamental to the normal development of tissues and organisms. In contrast, cell death by necrosis is usually accidental and therefore does not have such significance.

The role of apoptosis in cell population control during development has suggested that there are inherent cellular programmes that lead the cell to self-destruct. This has been confirmed in a number of instances; e.g. in a small nematode, *Caenorhabditis elegans* (*C. elegans*), where each individual cell can be recognized, it has been found that in the hermaphrodite form of the worm the same set of 113 cells is destined for programmed cell death during embryogenesis, and another set of 18 cells later in life, for a total of 131 cells (1). Also, inhibition of RNA or protein synthesis can, in many cases, abrogate cell death by apoptosis (2), although it usually accelerates necrosis. Thus, it appears that gene expression is necessary for cell death. Yet, there is another level of complexity, as, in some instances, inhibition of protein or RNA synthesis, or even explusion of nuclei, does not prevent what otherwise appears to be programmed cell death (3). Such cells are thought to be primed for apoptosis.

The original use of the term apoptosis was primarily descriptive of the cellular morphology of dying cells (4). Although in the current literature most authors blur the precision of this term (3), it is still tenable to define apoptosis as cell death that differs from necrosis on a morphological basis, observable by light or by electron microscopy. The key features originally described included shrinkage and blebbing of the cytoplasm; preservation of the structure of cellular organelles, including the mitochondria; and condensation and margination of chromatin, although not all of these are seen in all cell types (*Figure 1*). It is generally assumed that these morphological changes result from a developmental programme for cell death that can be triggered by deprivation of a growth factor, or by addition of a xenobiotic compound such as a cancer therapeutic drug. The morphological criteria are still the most important when complex cell populations, such as tissues, are examined,

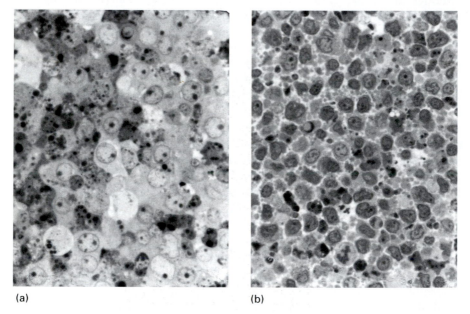

(a)   (b)

Figure 1. An illustration of the light microscopic appearance of apoptotic cells and their modification by a differentiation-inducing agent. (a) HL60 cells were exposed to calcium ionophore A23187 (10 μM for 8 h), embedded in epon, and 10 μm sections were stained with toluidine blue. Note the densely stained fragments of chromatin in the nuclei and cytoplasm of most cells. (b) HL60 cells treated as in (a), but first exposed to 1,25-dihydroxyvitamin D_3 (10^{-8} M for 48 h), which protects HL60 cells against apoptosis (10). Note the smaller (differentiated) cells, only a few of which show apoptotic nuclei.

and overall cell shrinkage and nuclear condensation are the easiest to recognize. These are presented in detail in Chapter 2, with special emphasis on the detection and quantitation of apoptosis *in vivo*, since this is a much more challenging task than recognition of apoptosis in tissue culture.

In pure cell populations, biochemical changes in chromatin and DNA degradation provide useful and often quantifiable means of detecting apoptosis. It is often forgotten, however, that random DNA degradation is not a specific test for apoptosis but simply demonstrates cell death. Although detection of DNA degradation may be useful as an adjunct method of quantitation, occurrence of apoptosis has be shown by morphological or by more specific biochemical methods.

The classical biochemical method for demonstrating apoptosis is the presence of oligonucleosome-sized fragments of DNA, which, when run on agarose gels, produce 'ladders', as discussed in Chapter 3 and illustrated in *Figure 2* (5, 6). It has been shown also that an earlier endonucleolytic cleavage of chromatin produces DNA fragments from 300 kb down to 50 kb in size (7, 8). Also, the observation that mitochondrial DNA is intact in early stages of apoptosis

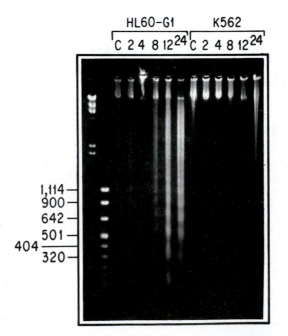

Figure 2. DNA ladder formation in HL60-G1 cells (a subclone of human leukaemia HL60 cells), but not in K562 human leukaemia cells, following exposure to doxorubicin (5 μM for HL60 cells and 10 μM for K562 cells, both for 24 h). DNA was extracted and run on 2% agarose gels and stained with ethidium bromide. DNA ladders indicative of nucleosomal DNA fragmentation became apparent after 8 h, concident with morphological appearances due to apoptosis (not shown). Microscopic examination of doxorubicin-treated K562 cells showed that these cells became necrotic (not shown).

provides a basis for a method which can detect and quantify apoptosis (9), as illustrated in *Fig 7* and discussed further in Section 6. These methods prove that apoptosis has occurred, although sometimes a lag period of several hours is necessary for the signs of apoptosis to become detectable, as will be discussed later in this chapter. It is possible that this lag can explain some situations in which morphological apoptosis is not accompanied by DNA fragmentation.

Analysis of all aspects of cell death has been greatly aided by application of flow cytometry (see Chapter 4). The instrumentation is now widely available in most academic or industrial centres, and can provide quantifiable data for practically every method of detection of apoptosis described in this volume. In the author's laboratory, determination of the sub-G1/G0 cellular DNA content of propidium iodide-stained cells (see Chapter 4, *Protocol 4*) has been found to be excellent for routine use in studies of the effects of chemotherapeutic drugs on cultured cancer cells (e.g. ref. 10). A more specialized and less accessible instrument, the laser-scanning cytometer, combines the advantages of flow cytometry with image analysis, providing information on cell morphology, and, if necessary, tissue architecture. The recent modifications of flow cytometric procedures for laser-scanning cytometry presented in Chapter 4 may prove particularly valuable for groups with access to this instrument.

The cascades which signal and execute cell death by apoptosis can be initiated by internal cues or by agents present in the extracellular environment. The cell death receptors described in Chapter 5 participate in the T cell-mediated cytotoxicity of target cells, and serve to illustrate the procedures used to study receptor-mediated pathways to apoptosis. Chapter 5 also presents an overview of pathways to cell-mediated cytotoxicity and methods which can distinguish cytoplasmic from nuclear manifestations of apoptosis.

Induction of apoptotic pathways by extracellular agents can also occur by membrane events that include liberation of sphingolipids, which act as messengers of cell death. Chapter 6 describes the sphingomyelin cycle in which a stress signal activates a cell membrane-associated enzyme, sphingomyelinase, resulting in the formation of ceramide from sphingomyelin. In this way various cytokines, such as TNF-α, interleukin-1, and γ-interferon, chemotherapeutic agents, and serum starvation can induce ceramide formation, which activates cellular protein kinases and protein phosphatases involved in cellular life and death decisions.

A well-known event in cell death is exteriorization of phosphatidylserine on plasma membrane. This change allows binding of the anticoagulant protein annexin V to this negatively charged phospholipid with great affinity, but is not entirely specific for apoptosis. Approaches that allow the distinction of apoptosis from necrosis based on annexin V techniques are the principal focus of Chapter 7, which also discusses other cytoskeletal and nucleoskeletal alterations associated with apoptosis.

Signals for apoptosis generated within the cell do not appear to be well understood, but are known to include metabolic alterations in the mitochondria that result in the disruption of mitochondrial transmembrane potential and changes in the cellular redox state. Chapter 8 addresses the mitochondrial permeability transitions and the role of various aspects of oxidative stress in the process of apoptosis. In particular, determination of peroxide and superoxide levels and cellular glutathione content are presented, to illustrate how surrogate end-points can be used to investigate and quantitate apoptosis-related phenomena.

An intermediate level of co-ordination of cell death versus survival signals is largely controlled by members of the Bcl-2 family of protein. These are discussed in Chapter 9, with a wealth of techniques optimized in the Reed laboratory, while investigation of the functioning of the caspase cascade is considered in Chapter 10. Extensive tabulation of the properties of caspases and of polypeptides cleaved during apoptosis concludes Chapter 10 and this volume.

2. Historical perspective

It is not always realized that apoptosis affects cells one at a time. Indeed, this form of cell death had been recognized earlier, as 'single cell necrosis'. The original descriptions were reported over a hundred years ago by pathologists studying liver diseases, but referred to as 'hyaline' or 'acidophilic bodies'. For instance, Councilman described hyaline bodies in the livers of patients dying of yellow fever (11), and subsequent electron microscopic studies concluded that these structures, also often called 'Councilman bodies', represent the remains of single dead hepatocytes (12, 13). Until recently, they were stated to result from coagulative necrosis of single cells (14), but, interestingly, they were noted to be eventually phagocytosed by macrophages. When the description 'apoptosis' was introduced by Kerr *et al.* in 1972 (4), this form of cell death fitted right into the definition, and, indeed, this was experimentally confirmed (12).

In the strictest sense, apoptosis refers to manifestations of a process, a programme leading to cell death, recognizable by morphological or a variety of biochemical criteria, which are discussed in subsequent chapters. The process itself, cell death, cannot, of course, be seen, and is detected by a combination of its manifestations. Thus, the terms ' apoptosis' and 'programmed cell death' are not exactly equivalent, nor is the phrase 'physiological cell death', which generally implies a developmentally determined programme, as opposed to 'chemoaptosis', in which the programme is triggered by xenobiotics such as cancer chemotherapeutic drugs.

3. Distinction of apoptosis from other forms of cell death

The current literature contains many examples of loose use of the term 'apoptosis'. There are several forms of cell death, and in all of them nuclear *meaning*

DNA becomes degraded, at some point. As mentioned above, demonstration of DNA damage or release of products of DNA degradation is by itself insufficient to justify description of the phenomenon as apoptosis. The distinction is more than a semantic debate, since the concept behind the term 'apoptosis' is the existence of an inherent cellular programme, somewhat similar to the programmes which drive cell differentiation, whereas 'necrosis' results entirely from circumstances outside the cell. Other forms of cell death, e.g. mitotic cell death, are insufficiently characterized to be considered at this time as biologically distinct entities.

The distinction between apoptosis and necrosis can be made biochemically (discussed on pp. 12–15) but is also very clear on purely morphological grounds. As sketched in *Figure 3*, the apoptotic cell shrinks, nuclear chromatin undergoes marked condensation, and it is expelled from the cells as apoptotic bodies that are phagocytosed by neighbouring cells. In contrast, necrotic cells first increase their cellular water content and thus their volume, the nuclei lose the typical chromatin structure which is often seen as irregular clumping and/or dissolution, and the cell membrane ruptures, discharging the cellular contents into the environment. Many of the manifestations of necrosis appear to be due to the depletion of cellular ATP by the agents or conditions precipitating necrosis, and it is known that there are steps in the apoptotic cascade that require ATP or dATP (15). This may be an incompletely explored area for the study of how to distinguish controversial cases of cell death; for further discussion of this interesting topic the report by Eguchi *et al.* should be consulted (16).

The criteria currently most useful for distinguishing apoptosis from necrosis are listed in *Tables 1* and *2*. Importantly, there are also similarities between

Table 1. Morphological differences and similarities between apoptosis and necrosis

	Differences		Similarities or confounding variables
	Apoptosis	Necrosis	
1. Nuclei	Pyknosis and karyorrhexis (dense condensation of chromatin)	Karyolysis preceded by irregular chromatin clumping	Damage occurs in both
2. Cytoplasmic organelles	Morphologically intact	Disrupted	Secondary damage in apoptosis
3. Cell membrane	Apoptotic bodies, blebbing	Blebbing and loss of integrity	Changes seen in both
4. Cell volume	Cells shrink	Cells swell	There may be no detectable changes
5. In tissues	Single cells affected	Groups of cells affected	In epithelia superficial cells are apoptotic and in groups
6. Tissue response	None	Inflammation	—

6

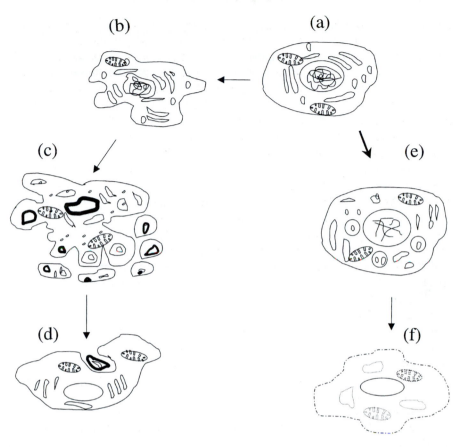

(b) (a)

(c) (e)

(d) (f)

Apoptosis

Necrosis

Figure 3. A sketch of the key morphological differences between apoptosis and necrosis. When a normal cell, depicted in (a), receives overriding signals to undergo apoptosis, it first exhibits an irregular contour and appears smaller (b). The chromatin then shows dense condensation, especially at the nuclear periphery, and small pieces of the cell, usually containing condensed chromatin, break off (c). The pieces, called apoptotic bodies, are taken up by phagocytosis by neighbouring cells, particularly macrophages if these happen to be present (d). The apoptotic bodies are then gradually digested by the phagocytic cells. In necrosis, the cells swell and chromatin often is alternatively diffuse or finely clumped (e). The cytoplasmic organelles, such as the mitochondria, may be swollen but remain intact. The necrotic cell eventually lyses, releasing all of its contents into the extracellular space, and thus eliciting an inflammatory response by the tissue.

apoptosis and necrosis, and, in view of these and a frequent overlap of characteristic features, conclusive evidence of the occurrence of apoptosis should demonstrate more than one morphological or biochemical criterion of apoptosis.

Table 2. Biochemical differences and similarities between apoptosis and necrosis

	Differences		Similarities or confounding variables
	Apoptosis	**Necrosis**	
1. Nuclear DNA damage	Nucleosomal and/or 50–300 kb fragments → ladders on gels	Random → smears on gels	Takes place in both, easier to detect in apoptosis
2. Nuclear gene expression	Usually needed	Not needed	Not needed in cells primed for apoptosis
3. Mitochondrial DNA damage	Spared	Occurs early	—
(a) DNases	Necessary	Not necessary	Lysosomal DNase and
(b) Proteases	Necessary		proteases are activated
(c) Transglutaminase	Frequent		in necrosis
5. Membrane function	Inact	Loss of function	—
6. Cell internal milieu			
(a) pH	Slightly acidic (pH 6.4)	Acidic	Both acidic
(b) Ca²⁺	Often increases	Always increases	Seen in both
(c) Na⁺/K⁺ pump	May be intact	Defective	—
(d) ATP	Required	Depleted	—

4. Apoptotic cascades

There appears to be an intricate but precisely ordered cellular machinery for self-destruction. This machinery is a subject of intense current investigations (discussed in several chapters in this volume) but some general principles have already emerged. First, the initiating signals can be either extracellular or intracellular. Secondly, many proteolytic enzymes are involved in apoptosis, which can be subdivided into classes which either initiate the process, propagate and amplify the signal, and those which attack the cellular structures to cause their collapse (*Figure 4*). Third, the mitochondria play an important role in this process, and many serve to integrate the various signals for apoptosis. Fourth, the pro-apoptotic machinery interacts with cellular survival mechanisms at several levels, including the mitochondria and the execution caspases (*Figure 5*). The suggested role for mitochondria in activation of the executioner caspase-3 is further illustrated in *Figure 6*.

5. Time course of apoptotic cascades

A wide range of times has been reported for the duration of apoptosis. The discrepancy is particularly marked when *in vivo* and tissue culture experiments are compared. For instance, *Figure 2* shows that DNA ladders, which signify a late stage of apoptosis, are evident at 8 h after addition of 5 μM doxorubicin to HL60 cells, and disruption of the inner mitochondria trans-

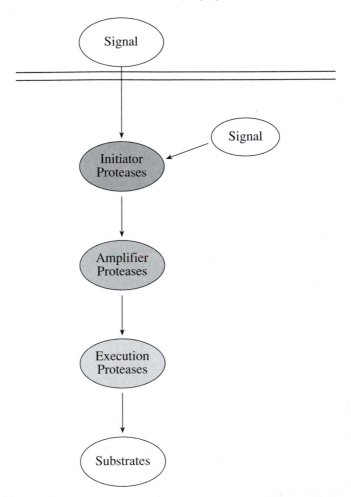

Figure 4. A conceptual outline of the major steps in the proteolytic cascade characteristic of apoptosis. Note that signals for apoptosis can be generated from the outside or the inside of a cell. The sequential activation of successive proteases (caspases) provides a fail-safe mechanism that makes it possible to abort a premature apoptotic signal, and also serves to amplify the signal to allow rapid finalization of an irrevocable decision to self-destruct.

membrane potential, an early event in apoptosis, can be detected in a similar *in vitro* system (human myeloma cells treated with a retinoid) within 1 h (17). In contrast, in an *in vivo* model of apoptosis that occurs in rat ventral prostate after castration, the process was reported to take 44 h, and more detailed analysis of this model suggested that the time gap between the apoptotic trigger and the appearance of minimally abnormal morphology was 12–16 h. The disassembly of apoptotic cells into apoptotic bodies was estimated to take

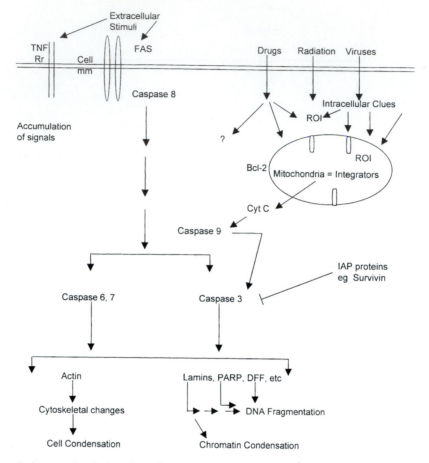

Figure 5. A more detailed outline of apoptosis-signalling pathways. The mitochondria are shown to act as the principal integrators of signals for apoptosis that do not have dedicated membrane receptors for that purpose. The survival signals which counter the pro-death signals are not shown, but components of the intracellular machinery that promotes survival are indicated (Bcl-2, survivin). A few examples of cellular targets for executioner caspases are also shown.

4–5 h, but the digestion of apoptotic bodies by the neighbouring cells, which phagocytose these remnants, appeared to be a slow process, requiring approximately 24 h (18).

Although these observations suggest that finding evidence of apoptosis in animal tissues allows a reasonable window of time, this situation is not always the case. Apoptotic bodies are frequently cleared from the tissues more rapidly than in the central prostate, and there is an additional uncertainty as to how rapidly the *in vivo* stimuli for apoptosis can reach the cell and be summated before they exceed the threshold for initiation of cell death pro-

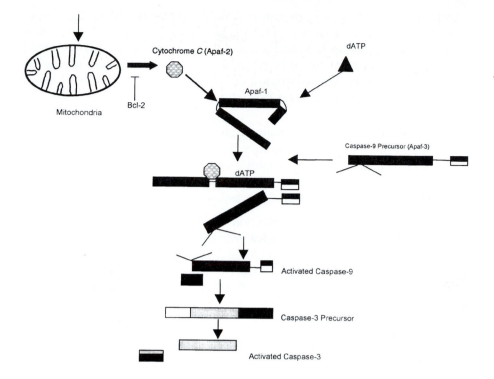

Figure 6. A detail of the role of mitochondria in triggering the caspase cascade. Note the release of cytochrome *c* and the requirement for dATP. (Modified from Li *et al.*, *Cell*, **91**, 479, 1997).

grammes. Thus, when looking for evidence of apoptosis in tissues, the old adage, 'the absence of proof is not necessarily the proof of absence', becomes particularly appropriate. The apoptotic cells and bodies may have been missed.

The importance of careful selection of the time parameters for detection of apoptotic is further discussed in Chapter 5.

6. Selection of methods

There are no infallible guidelines on how to chose a procedure that is ideal for the study of apoptosis and its manifestations. The first basic rule is to consider the objective of the study.

- If it is necessary to show simply that cells are dying, cell membrane permeability tests are sufficient and very convenient, e.g. trypan blue exclusion or permeability to propidium iodide.
- If it is to be demonstrated that apoptosis is the mode of cell death that has taken place, simple morphology gives an excellent indication of this process,

both *in vitro* and *in vivo*, though it is more fashionable to include more expensive procedures such as TUNEL or annexin V studies. When flow cytometric equipment is available for determination of sub-G1/G0 DNA content, this has been found to be exceedingly useful in daily use in the author's laboratory. Demonstration of DNA ladders, if it can be done, is very helpful in validating other procedures in the initial investigation of a new system.

• If the objective is to study mechanistic aspects of apoptosis, any of the techniques described in this volume may be appropriate, but *in vitro* systems and isolated components of apoptotic cascades are particularly important here. Examples are the release of cytochrome *c* from mitochondria and a demonstration of the cleavage of procaspase-3 (10).

The second basic rule for the selection of methods for the investigation of apoptosis is to decide which of the following is most important for the study: sensitivity, specificity, ability to quantitate, and speed/economy with which the determinations can be made. Sensitivity can be increased by enrichment for fractions containing dead cells, such as the floating cells in adherent cell populations, but, in general, these choices will vary depending on the circumstances of the individual laboratory, and careful review of the following chapters will provide guidelines for most investigators.

One procedure which is not described elsewhere, provides extraordinary specificity and reasonable quantitation in an *in vitro* setting, although it requires a significant amount of work and set-up for Southern analysis of DNA. The test depends on the established fact that although the mitochondria show functional aberrations early in apoptosis, the mitochondrial DNA remains intact, while in necrosis, mitochondrial DNA is subject to rapid degradation (9, 19).

6.1 Procedure for the determination of the increased mitochondrial to nuclear DNA ratio, for detection and quantitation of apoptosis

6.1.1 Principle

Mitochondria and other cellular organelles are preserved in apoptosis, but undergo rapid degradation in necrosis (4, 19). Mitochondria contain multiple copies of a 16.5 kb circular DNA genome which encodes several mitochondrial proteins, and mitochondrial ribosomal and transfer RNAs (20, 21). Nuclear DNA (nuDNA) is degraded in all forms of cell death. In contrast, mitochondrial DNA (mtDNA) remains relatively intact in apoptosis, but becomes degraded in necrosis (9, 19). Thus, determination of the integrity of any mitochondrial gene relative to the integrity of a nuclear gene provides an accurate procedure for distinguishing apoptosis form necrosis. In addition, the ratio of the signal intensity of the mitochondrial to the nuclear gene allows quantitation of the extent of apoptosis in the cell population (9).

DNA extracted from cells consists primarily of nuclear DNA (nuDNA) but also contains mitochondrial DNA (mtDNA), and the mixture can be subjected to restriction enzyme digestion and Southern blot analysis. The integrity of the mitochondrial DNA is determined by the ratio of the abundance of a mitochondrial gene representative of mtDNA (e.g. p72, ref. 22), to the abundance of a nuclear gene representative of nuDNA. An example of such a determination is shown in *Figure 7*, and its quantitation in *Table 3*.

Protocol 1.   Analysis of cell death by comparison of mitochondrial vs. nuclear DNA degradation assay

A. DNA isolation

Equipment and reagents

- Beckman J-6 centrifuge or equivalent
- digestion buffer: aqueous solution of 100 mM NaCl, 10 mM Tris–HCl (pH 8.0), 25 mM EDTA (pH 8.0), 0.5% SDS, 0.2 mg/ml proteinase K (PK)
- chloroform/isoamyl alcohol (24:1)
- phenol/chloroform/isoamyl alcohol (25:24:1) (Ameresco)
- ammonium acetate, 7.5 M
- ethanol, 100%, 70%
- TE buffer (pH 8.0): aqueous solution of 10 mM Tris–HCl, 5 mM EDTA

Method

1. Lyse the cells with the digestion buffer.

2. Incubate the lysed cells at 50°C for 12 h.

3. Extract the lysates once with phenol/chloroform/isoamyl alcohol, and twice with chloroform/isoamyl alcohol.

4. Precipitate the DNA by adding ammonium acetate to 2.5 M, mix well, then add 2 volumes of 100% ethanol.

5. Incubate at 4°C for at least 2 h.

6. Pellet the DNA by centrifugation at 4000 r.p.m. (12 000 *g*) for 30 min at RT.

7. Wash the pellets with 70% ethanol and repeat the centrifugation step.

8. Aspirate the supernatant and air-dry the pellet.

9. Dissolve the pellet in TE buffer.

10. Take an aliquot for spectrophotometric readings at 260 and 280 nm.

11. The DNA extract obtained as described here can be used for part B.

13

Protocol 1. *Continued*

B. Restriction enzyme digestion of extracted DNA, and Southern analysis

Equipment and reagents

- mt-gene probe (e.g. p72, 16S ribosomal RNA; see ref. 22)
- nu-gene probe (e.g. c-*myc*) (Oncor)
- *Eco*RI (Gibco BRL)
- *Hind*III (Gibco BRL)
- NaCl, 0.2 M

Method

1. Digest 100 µg of the DNA samples from part A with *Eco*RI or *Hind*III (3.5 U/µg DNA) for 8 h 37 °C.

2. Treat the reaction mixture with 2 units of DNase-free RNase for 1 h at 37 °C.

3. Extract the samples with organic solvents as described in part A (step 3).

4. Precipitate the digested DNA samples with 0.2 M NaCl and 100% ethanol, wash the pellets with 70% ethanol, aspirate the supernatant, and air-dry the pellet.

5. Dissolve the pellet in TE buffer and quantitate using a spectro-photometer.

6. Electrophorese 10 µg of the DNA sample as described in Chapter 3, Protocol 2.

7. Depurinate and denature the samples in the gel as described in ref. 23.

8. Transfer the DNA as described by Southern (24) and immobilize the DNA on the membrane by drying and baking at 80 °C for 2–5 h.

9. Nick translate the probes as described in ref. 25.

10. Add probes to the hybridization buffer.

11. Wash the membranes in SSC solution.

12. Wrap the membranes in plastic and expose to autoradiographic film at –80 °C for variable periods of time.

13. Analyse the film by densitometric image analysis for intensity.

7. Pitfalls

There is no doubt that apoptosis research will remain a vitally important area of biological and medical research for some time to come. This very fact provides some downsides, however. These include the sometimes exaggerated claims of specificity for methods or reagents, the profusion of reports that may be contradictory, and the expectation of some that the quality of the science

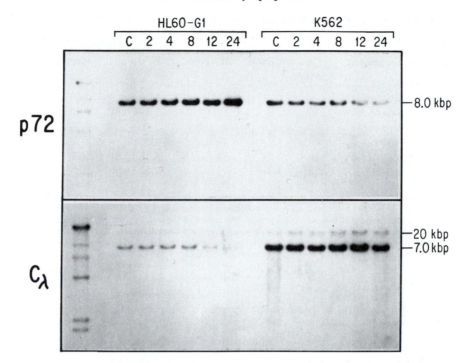

Figure 7. A Southern blot of the DNA shown in *Figure 2*, hybridized first to a mito-chondrial DNA probe (p72), then rehybridized to immunoglobulin lambda constant region gene (*C*). Note that the mitochondrial gene signal increases during doxorubicin-induced apoptosis of HL60 cells, but decreases during doxorubicin-induced necrosis of K562 cells. In contrast, the nuclear gene signal decreased during apoptosis of HL60 cells, but increased slightly during necrosis of K562 cells. The slight increase in the nuclear gene signal is due to the enrichment of the DNA sample with nuclear DNA because of the loss of mitochondrial DNA.

in a project or report may be elevated by the simple expedient of including an experiment or two on apoptosis. It is therefore probably more important than in any other field of science to maintain a highly critical attitude to assertions that a particular experimental manoeuver results in the occurrence of apoptosis.

The usual precaution is to demonstrate features of apoptosis by more than one independent approach. It is also important to prevent artefactual DNA damage that may mimic apoptotic DNA fragmentation. The loading of intact cells to be lysed in the gel is described in Chapter 3, Protocol 1. Overloading of gels with DNA produces smears rather than ladders, and attention to this point is a simple way to improve one's credibility. Another precaution often overlooked is to establish precisely the diploid DNA content of the cells under investigation before determining the sub-G1 values, since tumour cell populations are not infrequently composed of a mixture of diploid and

Table 3. Ratios of mtDNA to nuDNA during apoptosis in leukaemic cell lines

Cell line	Treatment	Concentration (mM)	Duration (h)	p72/c-*myc* ratio[a] EcoRI	Hind III
MOLT-4	Control	—	—	1.00	1.00
	Doxorubicin	10	12	3.43±0.56	2.50±0.51
	Dox	10	24	7.50±1.27	7.70±0.92
U937	Control	—	—	1.00	
	Teniposide	5	12	4.18±0.57	ND[b]
	ARA-C	10	12	4.24±0.11	ND

[a]Nuclear DNA and mtDNA levels were compared directly by sequential hybridization of the same membrane with probes for c-*myc* as a representative nuclear gene, and p72, a probe for mtDNA. Following densitometry of the autoradiographs, the ratios of the relative values of mtDNA/nuclear gene were obtained for untreated cells ('controls') and for each treated group. Control mtDNA/nuDNA ratios were converted to a value of 1.00, and the 'treated' mtDNA/nuDNA ratios were mutliplied by the same conversion factor yielding the values presented, ±S.E.M.
[b]Not done.

aneuploid cells. Under these circumstances the diploid peak can mimic the apoptotic sub-G1 peak.

The avoidance of these and other pitfalls and shortcuts should accelerate the acquisition of meaningful advances in apoptosis research.

Acknowledgements

I thank Dr Xuening Wang for help with the figures and Ms Claudine Marshall for secretarial assistance. The experimental work in my laboratory is supported by USPHS grant RO1-CA44722 from the National Cancer Institute.

References

1. Sulston, J.E., Schierenberg, E., White, J.G., and Thomson, J.N. (1983). *Dev. Biol.*, **100**, 64.
2. Wylie, A.H., Morris, R.G., Smith, A.L., and Dunlop, D. (1984). *J. Pathol.*, **142**, 67.
3. Eastman, A., Grant, S., Lock, R., Tritton, T., VanHouten, N., and Yuan, J. (1994). *Cancer Res.*, **54**, 2812.
4. Kerr, J.F.R., Wylie, A.H., and Currie, A.R. (1972). *Br. J. Cancer*, **26**, 239.
5. Skalka, M., Matyasova, J., and Cejkovan, M. (1976). *FEBS Lett.*, **72**, 271.
6. Wylie, A.H. (1980). *Nature*, **284**, 555.
7. Oberhamner, F., Wilson, J.W., Dive, C., Morris, I.D., Hickman, J.A., Wakeling, A.E., Walker, P.R., and Sikorska, M. (1993). *EMBO J.*, **12**, 3679.
8. Walker, R.P., Weaver, V.M., Lach, B., Leblanc, J., and Sikorska, M. (1994). *Exp. Cell Res.*, **213**, 100.

9. Tepper, C.G. and Studzinski, G.P. (1993). *J. Cell. Biochem.*, **52**, 352.
10. Wang, X. and Studzinski, G.P. (1997). *Exp. Cell Res.*, **235**, 210.
11. Councilman, W.T. (1890). In *Report on the etiology and prevention of yellow fever* (ed. G.M. Sternberg), United States Marine Hospital Service, Treasury Dept., Document No 1328 (Public Health Bulletin 2), pp. 151–159. Government Printing Office, Washington, D.C.
12. Kerr, J.F., Cooksley, W.G., Searle, J., Halliday, J.W., Halliday, W.J., Holder, L, Roberts, I., Burnett, W., and Powell, L.W. (1979). *Lancet*, **2**, 827.
13. Lacronique, V., Mignon, A., Fabre, M., Viollet, B., Rouquet, N., Molina,T., Porteu, A., Henrion, A., Bouscary, D., Varlet, P., Joulin, V., and Kahn, A. (1996). *Nat. Med.* **2**, 80.
14. Farber, J.L. and Rubin, E. (1988), in: *Pathology*, J.B. Lippincott Co., Philadelphia, p 15.
15. Liu, X., Kim, C.N., Yang, J., Jemmerson, R., and Wang, X. (1996). Cell, **86**, 147.
16. Eguchi, Y., Shimizu, S., and Tsujimoto, Y. (1997). *Cancer Res.*, **57**, 1835.
17. Marchetti, P., Schraen-Maschke, S., Thomas, A.M., Dhuiege, E., Belin, M.T., and Formstecher, P. (1999), In *Proceedings of the steroid receptor superfamily AACR special conference* (ed. M.G. Rosenfeld and C.K. Glass), AACR, Philadelphia, PA. p. A23.
18. Hu, Z., Ito, T., Yuri, K., Xie, C., Ozawa, H., and Kawata, M. (1998). Cell Tissue Res, **294**, 153.
19. Murgia, M., Pizzo, P., Sandonia, D., Zanovello, P., Rizzuto, R., and DiVirgillio, F. (1992). *J. Biol. Chem.*, **267**, 127.
20. Anderson, S., Bankier, A.T., Barrell, B.G., deBruijn, M.H.L., Coulson, A.R., Drouin, J., Eperon, I.C., Nierlich, D.P., Roe, B.A., Sanger, F., Schreire, P.H., Smith, A.J.H., Staden, R., and Young, I.G. (1981). *Nature*, **290**, 457.
21. Clayton, D.A. (1984). *Annu. Rev. Biochem.*, **53**, 573.
22. Tepper, C.G., Pater, M.M., Pater, A.M., Xu, H.M., and Studzinski, G.P. (1992). *Anal. Biochem.*, **203**, 127.
23. Wahl, G.M. Stern, M., and Stark, G.R. (1979). *Proc. Natl. Acad. Sci. USA*, **76**, 3683.
24. Southern, E.M. (1975). *J. Mol. Biol.*, **98**, 503.
25. Maniatis, T., Fritsch, E.F., and Sambrook, J. (1982). *Molecular cloning: a laboratory manual.* Cold Spring Harbor Lab., New York.

2

Morphological recognition of apoptotic cells

JAMES W. WILSON and CHRISTOPHER S. POTTEN

1. Introduction

In this chapter we outline the basic techniques that can be used to identify apoptotic cells on the basis of morphology. All these techniques require the use of some type of microscope, and obviously, the better the microscope then the easier your job is in assessing cell morphology. In addition to the subjective assessment of cellular/nuclear appearance, the use of immuno-cytochemical techniques specifically designed for the recognition of apoptotic cells is also outlined. Finally, we discuss how to extract quantitative informa-tion from such observations, and how such quantitative data may be inter-preted. Techniques for detecting the biochemical changes accompanying apoptosis are described in subsequent chapters.

1.1 Key morphological features of apoptotic cells

A seminal paper in apoptosis research was that by Kerr *et al.* (1). Their electron microscopic study of prednisolone-induced cell death in the kidney defined the term apoptosis and provided the standard reference for the key morphological features associated with this form of cell death. A cell under-going apoptosis proceeds through various stages of morphological change (see *Figure 1*). These are shrinkage of the cell away from its neighbours, plasma membrane blebbing, cytoplasmic and nuclear condensation, non-random cleavage of chromatin, margination of chromatin in the nucleus, nuclear fragmentation, and cellular fragmentation into smaller apoptotic bodies. Cells and cell fragments are ultimately phagocytosed by neighbouring cells and 'professional' phagocytes.

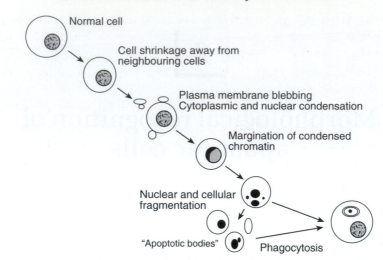

Normal cell

Cell shrinkage away from neighbouring cells

Plasma membrane blebbing
Cytoplasmic and nuclear condensation

Margination of condensed chromatin

Nuclear and cellular fragmentation

"Apoptotic bodies"

Phagocytosis

Figure 1. Schematic diagram of morphological changes associated with apoptosis.

2. Light and fluorescent microscopy techniques for the assessment of apoptosis

2.1 Preparation of cell or tissue samples

2.1.1 Slide preparation

The majority of protocols outlined in this chapter require the mounting of samples on glass microscope slides. In order to increase the adhesiveness of cell or tissue sections, slides need to be coated, or 'subbed'. The two most common subbing agents are gelatine and 3-aminopropyl triethoxysilane (APES). Gelatine is a good general purpose reagent, however, if microwave-based antigen retrieval is used for immunohistochemistry that is to be carried out in parallel with morphological assessment, then APES is recommended.

Protocol 1.   Gelatine subbing

Reagents
- 10% acetic acid
- ethanol (90, 95, 100%)

- 0.5 g gelatin

Method

1. Acid-wash the slides or coverslips for 1–2 days, in 10% acetic acid/90% ethanol.

2. Rinse (× 5) in deionized water, then 95% ethanol, and finally 100% ethanol. Store in 100% ethanol.

---

3. Dissolve 0.5 g gelatin in 100 ml of deionized water (at 65°C). Allow the solution to cool to room temperature prior to filtering through Whatman number 1 filter paper. Store the gelatine at 4°C.

4. When required, re-melt the gelatine using gentle heat (i.e. in a water bath).

5. Dip slides/coverslips in liquid gelatine, in a dust-free environment (i.e. a microbiological safety cabinet).

6. Place the slides/coverslips in a rack and allow to dry vertically.

Protocol 2.   APES subbing

Reagents
- decon-90
- 2% APES
- acetone

Method

1. Wash the slides in decon-90 (2% in water) for 5 min, and warm water for 5 min.

2. Place the slides in an acetone bath for 1 min, prior to immersion in 2% APES in acetone, for 5 min.

3. Rinse the slides in fresh acetone for 1 min, and in running deionized water for 5 min.

4. Place the slides at 60°C for 1 min to dry.

2.1.2 Preparation of tissue sections

Protocol 3.   Using formaldehyde-fixed, paraffin wax-embedded tissue

Equipment and reagents
- PBS, pH 7.4
- 4% formaldehyde
- chloroform
- ethanol
- microtome

Method

1. Wash the tissue (animal/human) in ice-cold, phosphate-buffered saline (PBS, pH 7.4), prior to fixation in 4% formaldehyde in PBS overnight, at 4°C.

2. Dehydrate the tissue through a series of graded alcohols (70% × 3,

Protocol 3. *Continued*

90%, 95%, 100% × 3). Allow 30–60 min for each step, depending on the size of the tissue sample.

3. Transfer the tissue to chloroform/ethanol. After 30–60 min, transfer to 100% chloroform. Change to fresh chloroform for a further 30–60 min.

4. Place the tissue in molten wax at 60°C, for 30 min. Place under partial vacuum and leave for a further 30 min. Change to fresh wax (at 60°C) and place under full vacuum for 2 h.

5. Dispense fresh wax into a mould, place on a cooling tray, and allow to begin to set.

6. Place the tissue in the required orientation within the wax and allow to set fully.

7. Section at 3–5 μm using a microtome. Expand the sections by floating on deionized water, at about 48–50°C. Pick up the sections on to APES- or gelatin-coated slides. Rack the slides and dry at 37°C overnight.

8. Store the slides in a dry, cool place. If slides are also to be used for immunohistochemistry, store at 4°C.

Protocol 4. Using frozen tissue sections

Equipment and reagents

- PBS
- liquid nitrogen
- OCT compound (optimal cutting temperature: OCT)
- cryostat and cryovials
- APES
- acetone
- methanol

Method

1. Wash the tissue in ice-cold PBS (pH 7.4). Remove excess moisture by blotting on a paper towel.

2. Freeze the tissue in the vapour phase of liquid nitrogen, by placing in a small petri dish and floating this on the surface of the liquid nitrogen. Store in cryovials, under liquid nitrogen, until use. For some tissues, i.e. skeletal muscle, freezing the tissue in liquid nitrogen-cooled isopentane is advised to give better preservation of tissue architecture.

3. Embed in OCT compound and section using a cryostat (tissue may be stored in cryovials at –80°C or lower, until required).

4. Pick up sections on to APES- or gelatin-coated microscope slides or coverslips.

5. Fix sections in acetone/methanol (1:1), for 3 min at –20°C, in a spark-proof freezer. Air-dry the sections for 10 min and store at –80°C until use.

2.1.3 Preparation of cell culture samples

The preparation of samples from cell cultures is relatively straightforward and less labour intensive than the preparation of tissue sections.

Protocol 5. Preparation of cells in suspension cultures

Method

1. Harvest the cells from tissue culture.

2. Resuspend the cells at a density of ~1 $\times$ 10^6 cells/ml.

3. Load 100–200 μl of cell suspension into the reservoir of a cytospin slide, and spin on to coated slides, at 500 r.p.m. (c. 30g) for 2–3 min (Shandon Cytospin 3). Allow to air-dry.

Cells may be fixed immediately after harvesting or after spun slides have been air-dried. Common fixation techniques are 30 min in 4% paraformaldehyde in PBS (pH 7.4) at 4°C or 3 min in acetone/methanol (1:1), at –20°C.
Unfixed cells in suspension may also be mixed directly with nuclear stains, as detailed in *Protocol 10.*

Protocol 6. Preparation of cells from monolayer cultures

Method

1. Either, harvest cells using trypsin/EDTA treatment and treat as in *Protocol 5.*

2. Or, if cells are able, grow them on glass coverslips (or slides). Coverslips can be easily removed from the tissue culture dish using a suitable implement. Cells on slides/coverslips can be fixed according to the methods in *Protocol 5.*

3. Alternatively, grow cells in chamber slides. Fixation of cells on chamber slides should be carried out using 100% methanol at –20°C; acetone will dissolve the plastic slide. As an alternative to chamber slides, cells can be grown in conventional culture flasks, and when assessment of morphology is required the top and side of the flask may be removed by a model-makers electric cutting tool, or by other means.

2.2 Nuclear counterstains

There are numerous stains and dyes that are employed to assess nuclear morphology. This chapter will concentrate on the methods we currently employ in our laboratory, which we feel allow accurate assessment of nuclear morphology and apoptosis. Stains and dyes are separated according to their use in either light or fluorescent microscopy.

2.2.1 Stains for light microscopy

Haematoxylin and thionin blue are the two nuclear counterstains that are most frequently used in our laboratory. They are both used for conventional light microscopic examination of sections of formaldehyde-fixed, wax-embedded tissue. They give a blue stain to all nuclei and have good contrast. The condensed chromatin within apoptotic cells stains particularly heavily. Mitotic nuclei also stain darkly, but can be differentiated because of their larger size and more fuzzy appearance. The chromatin masses within apoptotic cells tend to have 'sharp' borders. In the intestinal epithelium, mitotic cells tend to appear more displaced towards the centre of the crypt lumen, although this can also be true of some apoptotic cells/bodies. In our laboratory we routinely use haematoxylin and eosin staining for assessment of apoptosis in tissue sections, in parallel with whole-tissue autoradiography for studying tritiated thymidine incorporation. Thionin blue is most commonly used as a counterstain when apoptosis is being assessed in parallel with immunoreactivity in wax-embedded tissue sections. Examples of apoptotic cells in haematoxylin- and thionin-stained small intestinal epithelia are shown in *Figure 2*.

Wax-embedded tissue sections need to be hydrated prior to staining, as in *Protocol 7*. If the staining is carried out at the end of a procedure such as

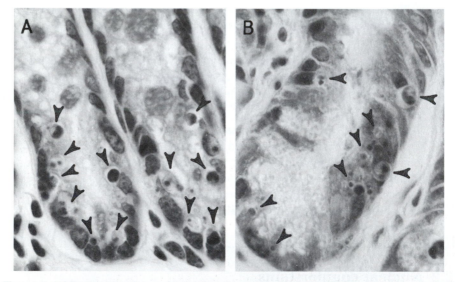

Figure 2. H&E staining of large intestinal crypt (**A**) and thionin blue staining of small intestinal crypt (**B**), from a mouse, 4 h after exposure to 16 Gy γ-radiation. Apoptotic cells/bodies are indicated by arrow heads.

immunocytochemistry or autoradiography, the sections will already be in a hydrated state and can be used directly, as detailed in *Protocol 8*.

Protocol 7. Rehydration of slides

Reagents
- xylene
- ethanol

Method
1. Warm the slides to 60°C for 10–15 min in an oven, to melt the wax.
2. Place in fresh xylene for 5 min, with constant agitation.
3. Transfer the slides to absolute alcohol for 5 min.
4. Rehydrate the slides through a graded series of alcohols: three further changes of 100%, then 95%, 90%, 70%, and 40%, with 3 min in each.
5. Finally, rinse the slides in deionized water.

Protocol 8. Haematoxylin (and eosin: H&E)

Reagents
- Gill's haematoxylin
- ammonia solution
- eosin
- ethanol
- xylene

Method
1. Place slides in Gill's × 2 haematoxylin for 3 min.
2. Rinse slides in running water for 1 min.
3. Give slides three dips in alkali water (deionized water with 4–5 drops of ammonia solution) and return to running water for a further minute. This results in the pink staining turning blue. If the blue coloration is too dark (this can be checked quickly using a microscope), it can hinder morphological assessment and the scoring of any parallel autoradiography. The stain may be lightened by placing slides in acid water (deionized water with a few drops of HCl) for a few seconds. Then rinse the slide in running water and re-examine.
4. Transfer to alcoholic eosin (0.4% eosin in 70% ethanol) for 1 min.
5. Rinse in running water for 1 min.
6. Dehydrate through a series of graded alcohols (40%, 70%, 90%, 95%, and 3–4 changes of 100% alcohol), with 5 min in each.
7. Place in xylene for 30 min, then mount slides using a permanent mount (XAM, DPX).

Protocol 9. *Continued*

8. Allow slides to dry overnight, prior to microscopic examination of sections.

Gill's haematoxylin no. 2 (product code 6765007, Shandon Inc.) is the stain of choice in our laboratory as it is relatively stable and gives reproducible and uniform staining. There are other, commomly used haematoxylins, including Ehrlich's, Meyer's, and Harris's. Ehrlich's stain requires two months to ripen prior to use, although it is very stable and gives excellent morphology. Meyer's stain is prepared with chloral hydrate, and Harris's with mercuric oxide, which make them unattractive for general laboratory use. Both the latter stains go off quickly compared with Gill's. More information regarding the applications of different haematoxylin stains can be obtained in: *Theory and practice of histological hechniques* (ed. J.D. Bancroft and A. Stevens), p. 107. Churchill Livingstone, Edinburgh (1990).

Protocol 9. Thionin blue

Reagents

- Thionin blue (4 parts solution A and 1 part solution B)
- 80% methanol
- solution A: 0.5 g thionin acetate (Sigma, T7029) in 100 ml methanol (filtered)
- solution B: 8 ml glacial acetic acid, 18 ml 5 M sodium hydroxide, made up to 100 ml with deionized water
- 100% ethanol

Method

1. Place the slides in thionin blue for 10 min.

2. Transfer to 80% methanol for 10 min.

3. Give the slides 10 dips in 95% ethanol. Repeat.

4. Transfer to absolute ethanol for 10 min.

5. Treat as for *Protocol 8*, steps 7 and 8.

Thionin may be reused many times before staining intensity is impaired. Methanol (80%) may also be reused, until it becomes too discoloured.

2.2.2 Stains for fluorescence microscopy

Fluorescent nuclear counterstains are most appropriate for use with cultured cell systems, and when fluorescent detection methods are being used in parallel for immunohistochemistry. The most commonly used are 4′,6-diamidino-2-phenylindole (DAPI), Hoechst 33258 and 33342, acridine orange, and propidium iodide. As with the stains used for light microscopy, the tissue must be hydrated prior to staining.

Hoechst 33258 stains non-apoptotic human and murine nuclei differentially. Human nuclei have a uniform, diffuse stain, whereas murine nuclei demonstrate several small, brightly staining bodies (2). Although there are no data to suggest that apoptotic cells from different species are stained differentially, this property of the Hoechst dye can be very useful when carrying out morphological assessment of xenografted human tissues in immune-deficient mice (3, 4), as it permits a distinction to be made between mouse and human tissue. One possible caveat with the use of Hoechst dyes is that they have been shown to induce apoptosis (5). However, this is only going to be a problem in unfixed tissue, and the length of time required to induce this effect (3 h) means that it should be irrelevant in all but exceptional circumstances.

Protocol 10. Fluorescent stains

Reagents

Make up stains as follows:
- DAPI: 5 mg/ml stock in methanol. Prior to use, dilute 1:10 000 in PBS, pH 7.4 (Sigma D9542).
- Hoechst 33258 and 33342: 100 μg/ml stock in PBS, pH 7.4. Prior to use, dilute 1:10 in PBS (Sigma B2883 and B2261).
- Acridine orange: 2 mg/ml stock in PBS, pH7.4. Prior to use dilute 1:400 in PBS (Sigma)
- Propidium iodide: 2 mg/ml stock in PBS, pH 7.4. Prior to use, dilute 1:1500 in PBS (Sigma P4170).

Method

1. Incubate the cells/sections with stain for 3–5 min. For cell suspensions, resuspend the cells at a density of ~2 × 10⁶ cells/ml and mix with an equal volume of dye.

2. Wash the cells/sections twice in PBS.

3. Mount using an aqueous-based mountant, with an anti-fade additive (Vectorshield, H1000; Vector Labs Inc.).

Acridine orange has been the dye of choice for many years for researchers studying developmental cell death in *Drosophila* (6–8). Acridine orange is a vital dye, i.e. it is excluded by viable cells. Staining of non-fixed, whole embryos with acridine orange allows the investigator to examine the spatial and temporal aspects of cell death in developing *Drosophila* embryos, using confocal fluorescence microscopy. Experimental details concerning this technique can be found in refs 6–8.

3. Electron microscopic techniques

Electron microscopy was the original technique used by Kerr *et al.* (1) in their seminal publication on apoptosis. It provides the most detailed information for the assessment of cell morphology, and hence the most accurate determination of apoptosis in tissues. However, it requires more expensive equipment and takes longer than other methods. *Protocol 11* describes a method for transmission electron microscopic study of cell lines and isolated cells from tissues. Prior to starting *Protocol 11*, cultured cells must be harvested by conventional means. Cells from normal tissues and tumours may be isolated by a variety of methods, including scraping and mechanical disaggregation, in combination with mild enzymatic treatment or chelating agents. Readers are advised to seek more specialized texts for the optimal preparation of specific tissues. Particular care needs to be taken when preparing samples for electron microscopy that are also going to be used for immunolabelling of proteins and RNAs. Here, the use of low-temperature embedding procedures is recommended in order to preserve antigenicity (9, 10).

Protocol 11. Preparation of cells for transmission electron microscopy (TEM)

Reagents
- 0.1 M sodium cocodylate buffer (pH 7.2–7.4)
- 2% paraformaldehyde
- glutaraldehyde
- 1% osmium tetroxide
- 2% magnesium uranyl acetate
- Spurr's resin mixture
- 0.3% lead citrate

Method

11. Resuspend the cells at a density of ~2 × 10^7/ml in 0.1 M sodium cacodylate buffer. For this protocol a 1.5 ml plastic centrifuge tube is a useful size to carry out the procedures in.

12. After 5 min, pellet the cells by centrifugation, and resuspend in 2% paraformaldehyde/2% glutaraldehyde in 0.1 M sodium cacodylate buffer (Tooze fixative). Leave the cells in the Tooze fix for 1 h.

13. Pellet the cells by centrifugation, and resuspend in 0.1 M sodium cacodylate buffer.

14. Pellet the cells again after a further 5 min, and resuspend in 1% osmium tetroxide, in 0.1 M sodium cacodylate buffer. Leave for 1 h.

15. After 1 h, pellet and resuspend the cells in 2% magnesium uranyl acetate (in 70% ethanol). Leave overnight, in the dark.

16. The following day, dehydrate the cells in a graded series of alcohols (3 × 70%, 90%, 95%), allowing 15 min in each.

17. Carry out three further changes in 100% ethanol, allowing 1 h between each change.

18. After dehydration, allow 1–2 h for infiltration of absolute alcohol/Spurr's resin mixture (1:1).

19. Change the alcohol/resin mixture for 100% Spurr's resin and allow further infiltration overnight.

10. The next day carry out a further three changes of resin, and finally polymerize the block overnight at 60°C.

11. Cut the sections (at 30–50 nm thickness) on to water using an ultramicrotome.

12. Expand the sections with chloroform, prior to transferring to copper mesh support grids. Allow the sections to dry.

13. Stain the sections with 2% uranyl acetate (in 70% ethanol) for 20 min.

14. Wash the sections three times with deionized water, prior to staining in 0.3% lead citrate (lead nitrate and lead acetate are also used routinely) for 3–4 min.

15. Wash three times in deionized water and allow to air-dry.

16. Examine the sections using a transmission electron microscope.

Electron microscopic (EM) evidence of apoptosis is illustarted in *Figure 3*.

4. Quantitation of apoptotic events

4.1 Methods

This is a very labour-intensive task, requiring much time being spent on the microscopic examination of prepared samples. Cells in suspension, stained with fluorescent nuclear dyes, are loaded on a haemocytometer and viable

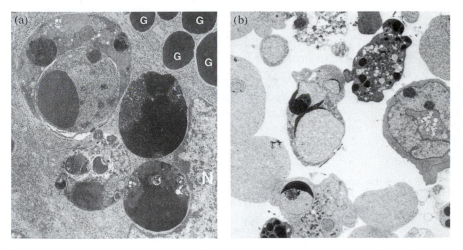

Figure 3. (a) Transmission electron micrograph of several apoptotic bodies which have been phagocytosed by a neighbouring cell. One of the phagocytosed bodies appears to be a cell which itself has previously phagocytosed another cell. The micrograph is of the base of a small intestinal crypt. The elecron-opaque bodies marked **G** are Paneth cell granules. To the bottom right of the picture is the nucleus of the phagocytosing cell (**N**). We are grateful to Dr T.D. Allen for the production of this electron micrograph. (b) Budding apoptotic bodies and crescents in HL60 cells treated with 20 μM etoposide for 6 h (supplied by G.P. Studzinski).

cells and apoptotic cells are logged simply by use of a tally counter. Cytospun slides, cell monolayers, and tissue sections may be scored in a similar way, except that a graticule can be placed in the eye piece of the microscope to substitute for the haemocytometer.

Methods do exist to make counting easier and to reduce the time spent staring down a microscope, which include flow cytometry (see Chapter 4). In our laboratory we use a Zeiss Axiohome system. This consists of a microscope linked to a computer, which can monitor the co-ordinate position of the microscope stage and allows the user to overlay a projected image of tally symbols on the counted cells. The microscope automatically keeps a record of the total number of each symbol used. This system is particularly useful when scoring tissue sections with little or no structural organization, i.e. tumours and hair follicles. When scoring, it is the general practice to record observations of 1000 cells from 10 random fields. Video capture of slide images is one method which allows the observer more independence from the microscope. Digital images can then be analysed on screen; however, this does not allow the observer to focus up and down on an object in the slide that may be slightly out of plane (focus), which can be critical for assessing the number of small apoptotic bodies. The use of serial images from confocal microscopy is desirable, since it eliminates this problem.

For structurally organized tissue such as the intestine, it is possible to score the cells on a positional basis. This technique has revealed important biological differences between cells and allowed a number of issues to be addressed. One critical procedure to carry out this scoring is the optimal preparation and orientation of tissue sections. The method used in our laboratory is outlined in *Protocol 12*.

Protocol 12. Preparation of tissue for the positional scoring of apoptotic events in gastrointestinal epithelia of mice

Method

1. Excise the small and large intestine and flush with ice-cold PBS.
2. Separate the intestine into different regions, i.e. duodenum, jejunum, upper ileum, lower ileum, caecum, mid-colon/rectum.
3. Cut into 1 cm lengths.
4. 'Bundle' together up to 10 pieces of intestine:
    (a) cut an 8–10 cm length of micropore tape and form a loop by sticking the two ends together,
    (b) lay the pieces of intestine parallel to each other within the loop and bind together with micropore tape,
    (c) cut a 4–5 mm thick cross-sectional disc from the taped bundle.

---

5. Process these smaller bundles for histology, as described in *Protocol 3*.

6. Cut sections through the bundles and process by one of the protocols described in Section 2.2.

7. Examine the slides.

If the tissue is prepared as described in *Protocol 12*, slides with good transverse sections of the intestine and good longitudinal crypt sections will be obtained. Cells can be numbered from the base of the crypt, starting at cell position one, with each side of a crypt (half-crypt) being scored separately. Fifty half-crypts are commonly scored for each region of the intestine; ideally from 4–5 bundles containing 5–10 individual cross-sections. The average length of a small intestinal half-crypt is 20 cells, i.e. 50 half-crypts is equivalent to 1000 cells. The data are recorded on a laptop computer, using custom-written software for data acquisition and analysis (PC Crypts, Steve Roberts, Bio-statistics and Computing Department, PICR). The recording of data in this way allows the frequency distributions of events (i.e. apoptosis, mitosis) to be plotted for each cell position. This is illustrated in *Figure 4*. All positions can be related to cell lineage position and, thus, stem cell properties can be studied (11, 12).

4.2 Problems in scoring apoptotic events

Quantitation of apoptotic events in cellular/tissue systems is always a topic that raises much debate. The major problem is whether the methods used accurately determine the frequency of apoptotic events, i.e. cell deaths, and how to interpret such numbers (11, 12). A number of factors can influence the frequency of apoptotic events, or 'apoptotic index'. These include:

(a) *Half-life*. The half-life of apoptotic cells differs between cell systems/tissues. In a tissue/cell line where apoptotic cells have a long half-life, samples at different times over a given period will show a higher frequency of events, compared with tissue where apoptotic cells have a short half-life, even though the absolute number of cell deaths may be the same.

Half-life itself is influenced by other factors. These include the rate at which cells are cleared by phagocytosis, migration of cells in a hierarchical system such as the intestinal tract, and the rate at which degradative changes take place. We have previously published data that suggest that the half-life of apoptotic cells may vary according to the cytotoxic agent used to induce apoptosis (see ref. 12 and references therein).

(b) *Section thickness*. Apoptotic cells are smaller than neighbouring, viable cells. They also fragment into yet smaller apoptotic bodies. Therefore, if very thin sections are cut through a tissue block, the apoptotic nuclei may

(a)

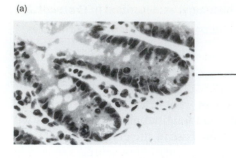

(b)

Viable Cell = 1
Apoptotic Cell = 2

Half-Crypts
#1: 12121121111121112121111111
#2: 121212121111111111111
etc...

(c)

Frequency of Apoptotic events

16
14
12
10
8
6
4
2
0

1 3 5 7 9 11 13 15 17 19

Cell Position

Figure 4. Scoring of apoptosis on a cell positional basis in the intestinal crypts. Small intestinal crypts from a mouse exposed to γ-radiation are shown in (a). Individual cells can be assigned a numerical value, according to their histological appearance. This is recorded on a laptop computer as a string of numbers, as demonstrated in (b). Pooling of data from 50 half-crypts can then be used to construct frequency distributions (c): the data shown are for irradiated crypts 4 h after exposure to 16 Gy γ-radiation (solid line) and unirradiated crypts (broken line). In addition to apoptosis, we commonly use this technique to analyse mitotic events, incorporation of tritiated thymidine, and the expression of nuclear antigens.

be 'missed', resulting in an underestimation of apoptotic frequency. The use of thicker sections may allow better visualization of apoptotic events, but overall the structural organization of the tissue may become less clear, as is the case with the intestinal tract. Whole-mount preparation of crypts can give better estimates of the frequency of apoptotic events (13, 14); however, their use is not suitable for routine examination of a large number of tissue samples.

(c) *Criteria selected for scoring.* It is important for observers to be consistent in their application of selection criteria for apoptotic cells. Most workers would tend to err on the side of caution and only score those cells that are unambiguously apoptotic. Apart from recognizing apoptotic cells, there is the question of whether one should count every apoptotic body or try to assess how many apoptotic events are represented by a group of apoptotic bodies. This situation can be illustarted by a comparison of the scoring of small intestinal epithelium and of proliferating hair follicles. The well-ordered structure of the small intestinal epithelium may allow the determination of whether given apoptotic bodies are the result of a single apoptotic event. In contrast, the hair follicle does not visibly demonstrate a highly ordered structure (although there is a cellular hierarchy), making this kind of assessment impossible. In this case, it is better to score all apoptotic bodies. As with the half-life of apoptotic cells/bodies, the number of apoptotic fragments arising from a single cell death may vary according to the cytotoxic insult.

(d) *Circadian rhythm*. This is certainly of relevance in the gastrointestinal tract, and probably also in other tissues, although diurnal variation in some tissues is less well characterized. We have shown that the rate of cell proliferation and cell migration in the intestinal crypts varies according to the time of day at which observations are made (15). When making comparisons between different treatments/experiments, therefore, it is essential that observations that are being contrasted are made at the same time of day.

(e) *Cellular hierarchies*. In hierarchical tissues, such as the intestinal epithelium, the sensitivity of any given cell to an apoptotic stimulus is determined by its position in that hierarchy. It is therefore useful to represent the frequency of apoptotic events in relation to this position within the hierarchy. This information can be very useful in determining whether effects are observed in selected cell types.

(f) *The denominator of the apoptotic index*. The apoptotic index (AI) is the number of apoptotic cells/bodies divided by the total number of cells. Within the well-ordered intestinal epithelium, the denominator can be determined reasonably accurately. For tissues with less visible organ-ization, i.e. tumours and hair follicles, cell counts will be less accurate. Also, in heterogeneous tissue such as tumours, there can be variation in the rates of cell proliferation and cell death in different regions of the tissue. Consequently, it can be useful to determine the ratio of the apoptotic index in relation to the proliferation index (cell in S phase, measured by bromodeoxyuridine or tritiated thymidine incorporation) or the mitotic index (MI). The AI:MI ratio effectively eliminates the denominator and any uncertainties associated with it. It also provides useful information on any net changes in cell number that are occurring within a tissue.

5. *In situ* detection of DNA strand breaks

In the last few years, techniques based on the detection of DNA strand breaks have gained popularity as methods for the identification of apoptotic cells. Two main techniques exist. They are: terminal deoxynucleotide transferase-mediated dUTP-biotin nick-end labelling (16), usually known by the acronym TUNEL; and DNA polymerase I-mediated *in situ* end-labelling, or ISEL (17, 18). The incorporation of biotinylated or digoxygenin-labelled bases into damaged DNA allows their subsequent detection by anti-biotin or anti-digoxygenin immunoglobulins. Conventional immunohistochemical detection techniques are then used. The TUNEL technique labels all DNA ends, i.e. 5′-overhangs, 3′-overhangs, and blunt ends. However, the ISEL technique only labels 3′-overhanging ends, making it less sensitive than TUNEL.

The techniques have been reported to be much more sensitive in detecting apoptotic cells, especially at an early stage, in comparison with morphological scoring of cells in H&E sections and flow cytometric analysis (19). For some tissues this technique seems to work quite well; however, we have found that the technique has many problems, as previously reported (14, 20). In our studies of radiation-induced apoptosis in the intestinal epithelium, we have obtained staining patterns using a modified TUNEL assay that were, in general, mostly consistent with observed morphological changes. However, a large number of cells that were obviously apoptotic, as assessed by morph-

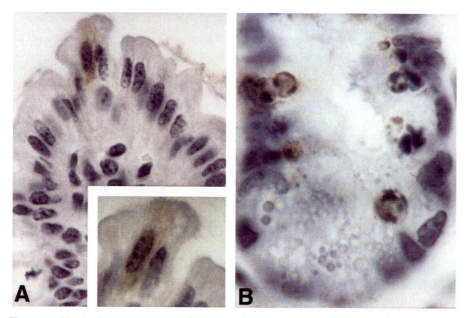

Figure 5. (A) Individual cell at the tip of a small intestinal villus, showing positive immunoreactivity using the TUNEL technique; the insert shows a high magnification view. Note that the cell does not demonstrate any visible changes in morphology that are commonly associated with apoptosis; however, alterations in the apical plasma membrane can be seen (see text). (B) Small intestinal crypt from an irradiated mouse, with positively stained apoptotic cells/bodies. These sections were stained with a modified version of the TUNEL technique, using the Oncor ApopTag® kit, as outlined in *Protocol 13*.

ology, were not stained. Previously published data showed that nearly 18% of false-negatives were obtained by TUNEL (14). Small alterations in TUNEL assay conditions, most notably in the concentration of the terminal deoxynucleotide transferase (TdT) and the duration of the proteolytic digestion, can result in the labelling of many non-apoptotic epithelial cells, especially on the villi.

The ISEL technique has been used to demonstrate that a small number of cells at the tip of the villi are undergoing apoptosis at any one time, prior to being shed into the intestinal lumen (21). We have been able to manipulate further the TUNEL assay conditions, to achieve a similar pattern of positively stained cells at the villus tip (as outlined in *Protocol 13*). It was also possible to observe abnormalities in the appearance of the brush border of these cells; a feature which has been demonstrated previously by scanning electron microscopy (22) and is associated with cells in the process of extrusion (see *Figure 5*).

Protocol 13.   TUNEL staining using the Oncor Apoptag® kit

N.B. The protocol described below is specifically optimized for sections of formalin-fixed, wax-embedded murine small intestine; we found it to be unsuitable for the study of human tumour samples. It is recommended that workers determine empirically the conditions required for optimal staining of their own tissue samples.

This technique follows the protocol in the ApopTag® kit handbook and uses the majority of the reagents supplied by the kit (product code S7100, Oncor). Listed below are the specific modifications we have made to the standard kit protocol.

Reagents

- proteinase K
- PBS
- 0.3% H_2O_2
- Oncor equilibration buffer
- Oncor reaction buffer
- Oncor stop-wash
- Oncor anti-digoxigenin peroxidase conjugate
- TdT

Method

11. Dewax and rehydrate sections as detailed in *Protocol 7*.

12. Incubate sections with proteinase K in PBS (40 μg/ml) for 20 min. N.B. This step is usually carried out on the bench. Day to day changes in room temperature can, therefore, have a significant effect on the degree of digestion and, hence, the final results. An ambient temperature of 18–20 °C is recommended.

13. Wash twice with PBS (5 min each wash), prior to incubation in 0.3% H_2O_2/PBS for 15 min.

Protocol 13. *Continued*

14. Wash twice with PBS (5 min each wash) and incubate with × 1 Oncor equilibration buffer for 30 min.

15. Tap-off the buffer and apply either 54 μl of working strength TdT (38 μl Oncor reaction buffer plus 16 μl TdT, as supplied) or 54 μl of control buffer (38 μl Oncor reaction buffer plus 16 μl distilled water) to each section. Coverslip and incubate for 1 h at 37°C.

16. Place slides in the Oncor stop-wash for 10 min, in a dish on a rocking table.

17. Wash in PBS for 5 min.

18. Incubate sections with × 1 Oncor anti-digoxigenin peroxidase conjugate for 1 h at room temperature.

19. Wash in PBS for 5 min prior to incubation in 3'3'-diaminobenzidine (DAB: 0.5 mg/ml) in 0.03% H_2O_2/PBS for 5 min.

10. Wash in PBS and counterstain with thionin blue, as described in *Protocol 9*.

See Oncor's web site at www.oncor.com.
N.B. This technique varies from the one previously published by Merritt *et al.* (14), which utilized non-Apoptag kit reagents.

Using an improved ISEL technique, Pompeiano *et al.* (23) have demonstrated that the majority of cells on the villi show positive staining, a pattern that we observed using the original TUNEL assay conditions (20). The fact that cells demonstrate DNA fragmentation prior to (in this case by 1–2 days) the morphological changes of apoptosis does not fit well with the current models of apoptotic events. It is very unlikely that, if DNA fragmentation has occurred in the villus cells, this is due to activation of the apoptotic endonuclease. DNA fragmentation *could* occur as part of the terminal differentiation of enterocytes. This low level of fragmentation would be detected by TUNEL (using the original assay condition (see ref. 20) and modified ISEL techniques (23). The less sensitive ISEL technique (21) and modified TUNEL technique (20) would be more specific in staining apoptotic cells, as they would only be capable of detecting the higher levels of DNA fragmentation associated with apoptosis, and not the low levels present in terminally differentiated cells.

Other workers have also reported caveats of the TUNEL assay. Biotinylated and digoxygenin-labelled nucleotides have been shown to bind to matrix vesicles in atherosclerotic plaques (24, 25). It was demonstrated that many of the nuclei labelled by the TUNEL assay were, in fact, transcriptionally active, as assayed by the expression of RNA splicing factor and the presence of polyA RNA tails (25). It is unlikely, therefore, that these were

otic cells, and, in addition, they did not show any morphological changes characteristic of apoptosis. The duration of proteolytic
treatment and dT concentration are clearly factors that have been found to
greatly influence results in our own and others' laboratories (25, 26). The
occurrence of DNA strand breaks in a non-apoptotic context has been a
matter of concern for other workers, and has been discussed in Chapter 7 of
Cell growth and apoptosis: a practical approach, Oxford Univerisity Press,
Oxford (1995). It has been reported that end-labelling techniques, when used
alone, cannot differentiate between apoptosis and necrosis (27). The
accumulation of strand breaks in post-mortem tissue prior to fixation could
also result in a number of false-positive events (28).

It would seem that it is best to approach end-labelling techniques with
caution. Always determine empirically what assay conditions give staining that
best matches the morphological criteria; always ensure that tissue is fixed as
rapidly as possible; and employ additional techniques, such as those to establish
transcriptional activity (25). The use of *in situ* hybridization for the identification of cell-type-specific markers, in parallel with TUNEL/ISEL techniques,
can also be useful when studying heterogeneous cell populations (29).

6. Other techniques

Other immunologically based techniques have also been developed, as com-
plementary methods for the identification of apoptotic cells. These include
the development of antibodies to single-strand DNA breaks (30); biotin-
labelled, hairpin oligonucleotides that bind to double-stranded DNA breaks
(31) and antibodies to caspase-cleavage products, such as the carboxyl termi-
nus fragment of actin (28). One of the most common techniques used is the
detection of phosphatidylserine in the outer leaflet of the plasma membrane
of apoptotic cells, using labelled annexin V. This technique is the subject of
Chapter 7. Here, it is sufficient to say that the technique can be used in conjunction with both light and electron microscopy. Tissue culture cells have
been mixed with labelled annexin V, prior to fixation and histological
examination (32). Annexin V has also been injected (intracardiac) into viable
embryos, which were subsequently fixed and sectioned for microscopy (33).

The discovery of other techniques for the identification of apoptotic cells
has been more serendipitous. Antibodies to both c-jun (34) and epidermal
growth factor receptor (35) have been shown to have specific immunoreactivity for apoptotic cells. Studies on nuclear pore protein regulation revealed that the nuclear pore protein POM121 was depleted from the nuclear
membrane in apoptotic cells (36). This loss could be monitored in cells transfected with a vector expressing a POM121–GFP (green fluorescent protein)
fusion protein, using fluorescence microscopy, and used to recognize
apoptotic cells.

7. Conclusions

Morphological assessment of apoptosis still has a key role in the study of cell death. The most important aspect of performing such analysis has to be the consistency of the investigator. It is always useful for anyone approaching the subject for the first time to have their scoring of apoptotic events checked by an investigator who has regular experience of the technique. There is no doubt that the recent development of DNA end-labelling and other techniques has helped in establishing the role of dysregulated apoptosis in various pathological conditions. However, caution must be urged in interpreting results obtained using these techniques—all that glitters is not gold. When contrasting data from different laboratories always be aware of what criteria and end-point the investigators have selected for the scoring of apoptotic cells.

Acknowledgements

Thanks to Dr Mike Bromley and Garry Ashton for informative discussions. Thanks to Dr Danielle Hargreaves and Nicola Crag for information on the modified TUNEL technique.

References

1. Kerr, J.F.R., Wyllie, A.H. and Currie, A.R. (1972). *Br. J. Cancer*, **26**, 239.
2. Cunha, G.R. and Vanderslice, K.D. (1984). *Stain Technol.*, **59**, 7.
3. Grimwood, R.E., Ferris, C.F., Nielsen, L.D., Huff, J.C. and Clark, R.A.F. (1986). *J. Invest. Dermatol.*, **87**, 42.
4. Kischer, C.W., Sheridan, D. and Pindur J. (1989). *Anatom. Rec.*, **225**, 189.
5. Zhang, X. and Kiechle, F.L. (1997). *Ann. Clin. Lab. Sci.*, **27**, 260.
6. Spreij, T.E. (1971). *Nethlds J. Zool.*, **21**, 221.
7. Abrams, J.M., White, K., Fessler, L.I. and Steller, H. (1993). *Development*, **117**, 29.
8. Hay, B.A., Wolff, T. and Rubin, G.M. (1994). *Development*, **120**, 2121.
9. Carlemalm, E., Villinger, W., Acetarin, J.D. and Kellenberger, E. (1985). In *Proc. 4th Pfefferkorn Conference 1985, Science of biological specimen preparation, SEM Inc.*, p. 147. AMF O'Hare, Chicago.
10. Hamilton, G., Hamilton, B. and Mallinger, R. (1992). *Histochem.*, **7**, 87.
11. Potten, C.S. (1995). In *Radiation and gut* (ed. C.S. Potten and J.H. Hendry), p. 61. Elsevier Science B.V., Amsterdam.
12. Potten, C.S. (1996). *Br. J. Cancer*, **74**, 1743.
13. Potten, C.S., Roberts, S.A., Chwalinski, S., Loeffler, M. and Paulus, U. (1988). *Cell Tissue Kinet.*, **21**, 231.
14. Merritt, A.J., Jones, L.S. and Potten, C.S. (1996). In *Techniques in apoptosis: a users guide* (ed. T.G. Cotter and S.J. Martin), p. 269. Portland Press Ltd, London.
15. Qiu, J.M., Roberts, S.A. and Potten, C.S. (1994). *Epith. Cell Biol.*, **3**, 137.
16. Gavrieli, Y., Sherman, Y. and Ben-Sasson, S.A. (1992). *J. Cell Biol.*, **119**, 493.

17. Ansari, B., Coates, P.J., Greenstein, B.D., and Hall, P.A. (1993). *J. Pathol.*, **170**, 1.
18. Wijsman, J.H., Jonker, R.R., Keijzer, R., Van de Velde, C.J.H., Cornelisse, C.J. and Van Dierendonck, J.H. (1993). *J. Histochem. Cytochem.*, **41**, 7.
19. Fujita, K., Ohyama, H. and Yamada, T. (1997). *Histochem. J.*, **29**, 823.
20. Merritt, A.J., Potten, C.S., Watson, A.J.M., Loh, D.Y., Nakayama, K., Nakayama, K. and Hickman, J.A. (1994). *J. Cell Sci.*, **108**, 2261.
21. Hall, P.A., Coates, P.J., Ansari, B. and Hopwood, D. (1994). *J. Cell Sci.*, **107**, 3569.
22. Potten, C.S. and Allen, T.D. (1977). *Ultrastruct. Res.*, **60**, 272.
23. Pompeiano, M., Hvala, M. and Chun, J. (1998). *Cell Death Different.*, **5**, 702.
24. Kockx, M.M., de Meyer, G.R.Y., Muhring, J., Bult H., Bultinck, J. and Herman, A. (1996). *Atherosclerosis*, **120**, 115
25. Kockx, M.M., Muhring, J., Knaapen, M.W.M. and de Meyer, G.R.Y. (1998). *Am. J. Pathol.*, **152**, 885.
26. Hegyi, L., Hardwick, S.J. and Mitchinson, M.J. (1997). *Am. J. Pathol.*, **105**, 371.
27. Mundle, S. and Raza, A. (1995). *Lab. Invest.*, **72**, 611.
28. Yang , F., Sun, X., Beech, W., Teter, B., Wu, S., Sigel, J., Vinters, H.V., Frautschy, S.A. and Cole, G.M. (1998). *Am. J. Pathol.*, **152**, 379.
29. Gochuico, B.R., Williams, M.C. and Fine, A. (1997). *Histochem. J.*, **29**, 413.
30. Naruse, I., Keino, H. and Kawarada, Y. (1994). *Histochem.*, **101**, 73.
31. Didenko, V.V., Tunstead, J.R. and Horsby, P.J. (1998). *Am. J. Pathol.*, **152**, 897.
32. Pellicciari, C., Bottone, M.G. and Biggiogera, M. (1997). *Eur. J. Histochem.*, **41**, 211.
33. Van-den-Eijnde, S.M., Boshart, L., Reutelingsperger, C.P., De-Zeeuw, C.I. and Vermeij-Keers, C. (1997). *Cell Death Different.*, **4**, 311.
34. Garrah, J.M., Bisby, M.A. and Rossiter, J.P. (1998). *Acta Neuropathol. Berl.*, **95**, 223.
35. Booth, C. and Potten, C.S. (1996). *Apoptosis*, **1**, 191.
36. Imreh, G., Beckman, M., Iverfeldt, K. and Hallberg-E. (1998). *Exptl Cell Res.*, **238**, 371.

3

Assessment of DNA damage in apoptosis

AKIRA YOSHIDA, RONG-GUANG SHAO, and YVES POMMIER

1. Introduction

Apoptosis is morphologically characterized by early cell shrinkage, chromatin condensation, nuclear fragmentation, cell surface blebbing, and membrane-bound apoptotic bodies (1) and these changes are described in Chapter 2. One of the most extensively studied biochemical events in apoptosis is chromatin fragmentation by endonuclease activation. DNA double-strand cleavage occurs in the linker regions between nucleosomes, and produces DNA fragments that are multiples of approximately 185 base pairs (bp) (2–4). These fragments can readily be demonstrated by agarose gel electrophoresis as DNA ladders, and this method has been widely used to detect and define apoptosis. In contrast, necrosis is accompanied by random DNA breakdown, with diffuse smear in agarose gels. The appearance of the nucleosomal DNA ladder in agarose gels is an important biochemical hallmark of apoptosis, although several investigators have observed apoptosis with no detectable internucleosomal DNA digestion (5, 6). In the absence of detectable inter-nucleosomal DNA fragmentation, high molecular weight fragments (HMW) ranging in size between 50 and 300 kb provide another characteristic feature of apoptosis (7, 8). In this chapter, we describe the commonly used methods to detect apoptotic internucleosomal and high molecular weight DNA frag-mentation and a sensitive filter elution method that can be used to quantitate these DNA fragmentations.

2. Types of DNA fragmentation

DNA fragmentation during apoptosis is considered to occur in two stages. Walker *et al.* (9) reported sequential degradation of genomic DNA, initially to high molecular weight (HMW) DNA fragments of approximately 300 kb, followed by the appearance of 50 kb loop-size chromatin fragments which were detected using pulse-field gel electrophoresis. Oligonucleosomal DNA fragments that produce the characteristic DNA ladder are released when the 50 kb fragments are further degraded. Certain cells, such as Molt-4 human T cell leukaemia cells (5) and DU145 human prostate cells (8, 10) do not display internucleosomal DNA fragmentation, although these cells die with a typical apoptotic morphology. In these cells, however, HMW DNA fragmentation can be observed, suggesting that HMW DNA cleavage is also an important biochemical marker of apoptosis and that internucleosomal DNA cleavage is dispensable for apoptosis (7, 11, 12). In general, haematopoietic cells, such as human leukaemia HL60 and Jurkat, are very sensitive to induction of apoptosis and undergo internucleosomal DNA fragmentation upon exposure to various apoptotic stimuli, including inhibitors of topoisomerase I (campto-thecin) or topoisomerase II (etoposide = VP-16) or staurosporine (13–17). On the other hand, epithelial cells (e.g. MCF7 human breast cancer or DU145 prostatic carcinoma cells) are generally more resistant to apoptosis (8, 10).

The presence of 5′-phosphate (5′-P) and 3′-hydroxyl (3′-OH) DNA termini in apoptotic cells is an important biochemical feature of apoptotic DNA fragments. Indeed, the terminal deoxynucleotidyl transferase-mediated dUTP–biotin nick end-labelling (TUNEL) method, which is commonly used to detect apoptotic cells, is based on the presence of 3′-OH termini (18). This method is described in Chapter 2. Another characteristic feature of apoptotic DNA fragmentation is that single-stranded DNA breaks are produced as well as double-strand breaks. Peitsch *et al.* (19) and Walker *et al.* (20) found that numerous single-strand breaks are generated in internucleosomal DNA regions during apoptosis. They proposed that DNA single-strand breaks accumulate in the linker regions between nucleosomes, eventually causing double-strand scissions which release oligonucleosomes.

Several nucleases are probably responsible for the DNA fragmentation in apoptosis (*Table 1*). Among the first reported were DNase I (21, 22) and Nuc18 (23). Nuc18 (cyclophilin) was isolated from apoptotic rat thymocytes (24). Recently, a 33 kDa DNase I-related nuclease (DNase Y) has been isolated from rat cells. Its gene encodes a 36 kDa protein that is ubiquitously expressed and is activated by proteolytic cleavage to a 33 kDa endonuclease (25). This nuclease resembles the DNase γ, a 34 kDa nuclease that was recently purified from rat thymocytes (26). Interestingly, DNase Y is not transcriptionally regulated in apoptosis (25). An acidic endonuclease, DNase II (27, 28) has also been proposed to be involved, as well as several Ca^{2+}/Mg^{2+}-dependent endonucleases (26, 29–31). However, it should be noted that DNase II produces 5′-OH and 3′-P DNA termini (32), which is not consistent with the characteristics of apoptotic DNA fragmentation. Two other Ca^{2+}/Mg^{2+}-dependent endonucleases have also been isolated from apoptotic rodent cells and another endonuclease was isolated from apoptotic mouse T cell hybridoma (30). It consists of three molecular forms (A, B, and C) with apparent molecular weights of 49, 47, and 45 kDa, respectively (30). In addition, a 97-kDa Ca^{2+}/Mg^{2+}-dependent nuclease producing internucleo-

Table 1. Overview of the apoptotic nucleases

Name	Ion dependency	MW (kDa)	Source	Reference
Nuc18	Ca^{2+}/Mg^{2+}	18	Rat thymocyte	23
ILCME	Ca^{2+}/Mg^{2+}	49, 47, 45	Mouse T cell hybridoma	30
(not named)	Ca^{2+}/Mg^{2+}	97	Rat hepatoma	31
(not named)	Ca^{2+}/Mg^{2+}	28	Human spleen	29
DNase I	Ca^{2+}/Mg^{2+}	31	Rat prostate	22
DNase γ	Ca^{2+}/Mg^{2+}	34	Rat thymocyte	26
DNase Y	Ca^{2+}/Mg^{2+}	33	Rat tissues	25
DNase II	(-)	40	Bovine spleen	28
CAD/DFF40	?	40	Rodent and human cells	33, 35
AN34	Mg2+	34	Human leukaemia HL60	37

somal DNA cleavage cells has been reported to be present in untreated rat hepatoma (31).

Very recently, a caspase-activated DNase (CAD) has been reported by S. Nagata and co-workers (33). CAD is a 40 kDa protein that induces internucleosomal DNA cleavage in isolated nuclei in the presence of Mg^{2+}. CAD activity does not require Ca^{2+}, and CAD forms a complex and co-purifies with an inhibitor of CAD (ICAD) that can be cleaved by caspase-3 during apoptosis (33, 34). CAD and ICAD have been isolated independently by X. Wang and co-workers who showed that DNA fragmentation and chromatin condensation depend on a heterodimeric protein composed of two DNA fragmentation factors of 45 and 40 kDa (DFF45 and DFF40 corresponding to ICAD and CAD, respectively) (35, 36). In our laboratory, we purified a 34 kDa Mg^{2+}-dependent and Ca^{2+}-independent endonuclease from etoposide-treated HL60 cells undergoing apoptosis (37), and designated this 34 kDa Mg^{2+}-dependent endonuclease, AN34 (apoptotic nuclease 34 kDa) (37). AN34 introduces single- and double-strand breaks in purified DNA and internucleosomal DNA cleavage in isolated nuclei (37). AN34-induced DNA breaks terminate with 3'-OH, consistent with characteristic products of apoptotic chromatin fragmentation (37). Interestingly, our data using caspase inhibitor (zVAD-fmk) suggest that the action of AN34 is also controlled by caspase (37), although AN34 is not inhibited by ICAD (DFF45) (A. Yoshida and Y. Pommier, unpublished data).

3. Measurement of high molecular weight DNA fragmentation by pulse-field gel electrophoresis

Since DNA degradation to oligonucleosomes is not essential for apoptosis and all cells must undergo DNA degradation to produce the large fragments (12), high molecular weight (HMW) DNA fragmentation has been considered a more reliable biochemical marker for apoptosis. When internucleosomal DNA cannot be demonstrated in some cells, there is considerable HMW

DNA fragmentation, producing fragments of 50–300 kb that can be identified by pulse-field gel electrophoresis.

Pulse-field gel electrophoresis is a specialized technique for resolving DNA molecules in the range of kilo- to mega-bases (Mb). By alternating the electric field between spatially distinct pairs of electrodes, HMW DNA fragments and chromosome-size DNA from 200 to over 12 000 kb can be separated because they are able to reorient and move differentially through the pores of an agarose gel.

3.1 Pulse-field gel electrophoresis procedures

We routinely use the CHEF-DR II or III pulse-field electrophoresis system (BioRad). The basic equipment is as described by the manufacturer. The CHEF-DR II system is an advanced pulse-field system based on the CHEF (clamped homogeneous electric fields) technique. The key to high resolution, sharp bands, straight lanes, and reproducible separations is a uniform electric field at all points of a gel, and an optimal 120 °angle of alternating pulses. The CHEF-DR II system accomplishes both of these by establishing the electric field along the contour of the hexagonal array of 24 electrodes. The basic unit of the CHEF-DR II system includes:

- gel chamber
- drive module
- pump
- Pulsewave 760 switcher
- model 200/2.0 power supply

The procedures used are as described in *Protocol 1*.

Protocol 1. Pulse-field gel electrophoresis procedures

Solutions

- TEN buffer: 5 mM Tris–HCl, 5 mM EDTA, 75 mM NaCl, pH 7.8.
- deproteinizing solution: 0.1 mg/ml proteinase K, 1% sarkosyl, 50 mM EDTA, 50 mM Tris–HCl, pH 7.8.
- plug-washing buffer: 10 mM Tris–HCl, 1 mM EDTA, pH 7.8.
- 0.5 × TBE running buffer: 45 mM Tris base, 45 mM boric acid, 1 mM EDTA, pH 8.2.

A. *Sample preparation in agarose blocks*

Any technique for the analysis of high molecular weight DNA fragmentation is critically dependent on its ability to isolate intact DNA from control cells and prevent further degradation of DNA from apoptotic cells. The technique of embedding cells in agarose plugs was evolved to solve this problem. Large DNA fragments are so fragile that they are broken by mechanical forces during their isolation. To prevent breakage of these

---

large DNA fragments, intact cells embedded in agarose are lysed and deproteinized. The agarose matrix protects the embedded DNA from shear forces and provides an easy way to manipulate samples. Processed agarose plug–DNA inserts are loaded directly into sample wells of agarose electrophoresis gels.

Method

1. Wash ~2 × 10^6 cells in phosphate-buffered saline (PBS).

2. Pellet the cells by centrifugation at 4°C for 10 min at 3000 r.p.m. (450 *g*).

3. Discard the supernatant and resuspend the cells with an equal volume of TEN buffer.

4. Prepare 1% low-melt preparative grade agarose (low melting point agarose) solution in TEN buffer by melting agarose in a microwave oven. Let the agarose solution cool to 50°C.

5. Add the 1% melted agarose to the cell suspension to give a final agarose concentration of 0.75%.

6. Pipette into mould chambers and allow cooling to 4°C for ~20 min.

7. The agarose plugs are removed from the mould and then incubated in deproteinizing solution at 45°C for 16 h.

8. Wash with plug-washing buffer once an hour, three times.

B. *Gel casting*

Method

1. Place the casting stand on a levelled surface.

2. Attach a comb to the holder and adjust the height of the comb to ~2 mm above the casting stand surface.

3. Prepare the desired concentration of agarose (see Section 3.2) in 0.5 × Tris–borate electrophoresis buffer.

4. Melt the agarose in a microwave oven.

5. Pour the molten agarose into the casting stand and allow to cool for 1 h at room temperature.

6. Carefully remove the comb holder and comb.

7. Sample plugs can be added to the wells while the electrophoresis gel remains in the casting stand.

C. *Loading the sample*

Method

1. Place the sample plugs on a smooth clean surface, and cut them to size. Samples should be less than 90% of the height of the wells.

Protocol 1. *Continued*

2. Fill each sample well with low-melt agarose at an agarose concentration equal to that of the gel, and allow the agarose to harden at room
   temperature for 15 min.

D. Running the gel

Method

1. Pour 0.5 × TBE buffer into the chamber.

2. Turn on the recirculating pump, and allow the buffer to equilibrate to
   the desired temperature (14°C is recommended).

3. Slide the gel on to the surface of the chamber. Check the buffer level
   so that the gel is covered by about 2 mm buffer.

4. Turn on the Pulsewave 760 switcher and power supply. Set the Pulsewave 760 parameters (the initial and final pulse time, ratio, running
   time) and voltage. For details of the conditions for electrophoretic
   separations see Section 3.2.

E. Staining the gel

Method

1. After electrophoresis, place the gel into 0.5 μg/ml ethidium bromide
   solution and let it stain for 20–30 min.

2. Destaining is performed in distilled water for 1–3 h.

3. The DNA can be visualized by placing the gel on a UV transilluminator
   (254–360 nm).

3.2 Strategies for electrophoretic separations

Several parameters must be considered before performing an electrophoretic
separation of HMW DNA. The separation of large DNA molecules in agarose
gels is affected by the agarose concentration, buffer concentration, temperature, pulse times, voltage, and total electrophoresis running time.

- The agarose concentration determines the size range of DNA molecules
  separated, and the sharpness or tightness of the bands. Agarose concentrations of 1% are useful in separating DNA molecules up to 3 Mb in size.

- Tris–borate buffer (TBE), at a concentration of 0.5 × and chilled to 14°C, is
  recommended for use in pulse-field electrophoresis.

- The migration rate of DNA molecules through agarose gel is determined by
  voltage, pulse time, and running time. The DNA migration velocity
  increases with increasing voltage and temperature, and decreases with
  increasing agarose concentration. Greater migration is accompanied by

Table 2. Run parameters

DNA size (Mb)	<0.1	0.1–0.2	2–4	>4
Agarose (%)	1.5	1–1.5	0.8–1	0.5–0.8
TBE buffer	0.5 ×	0.5 ×	0.5 ×	0.5 ×
Temperature (°C)	14	14	14	14
Voltage (V)	200	150–200	125–175	40–75
Pulse time (s)	1–10	1–10	1–10	10–60 min
Run time (h)	6–12	18–24	24–72	3–6 days

Table 3. Examples for pulse mode

DNA size (Mb)	0.2–23 kb	0.05–1	0.2–2.2	1–5	3.5–5.7
Mode	2	2	3	3	1
Agarose (%)	1.5	1	1	0.8	0.6
Pulse time (s)	1–4	50–90	60–90	120–240	60 min
Run time (h)	5	24	15–19	24–36	160
Voltage (V)	200	200	200	150	50

decreased band sharpness. Increased band sharpness requires longer run times at low voltage. Therefore, a compromise in voltage, temperature, and agarose concentration should be used to give the maximum allowable band sharpness within a reasonable run time. *Table 2* gives a general starting point for parameters to separate DNA molecules in the various size categories.

There are three common pulse modes used in CHEF analysis (see *Table 3*). Mode 1 is a single pulse time for the duration of the run. Mode 2 is an abrupt change in pulse time, with one pulse setting for the first part of the run and a second pulse setting for the remainder of the run. Mode 3 is a gradual linear change of the pulse time from the initial to the final pulse time.

4. Detection of internucleosomal DNA fragmentation by standard agarose gel electrophoresis

Internucleosomal DNA fragments can readily be detected by agarose gel electrophoresis. The small DNA fragments that are isolated from apoptotic cells are well resolved by gels. The simple and reproducible detection of internucleosomal DNA fragmentation has made it a biochemical hallmark of apoptosis. However, it should be noted that some cell lines (e.g. MCF-7, DU145) show almost no DNA ladders, although they show the typical morphological changes of apoptosis (see Section 1).

Protocol 2. DNA extraction and standard agarose gel electrophoresis

Equipment and reagents
- horizontal agarose gel electrophoresis chamber
- electric power supply
- sample comb
- ethidium bromide solution
- UV transilluminator
- TAE (40 mM Tris–acetate, 1 mM EDTA) or TBE buffer (45 mM Tris–borate, 1 mM EDTA)
- DNA sample loading solution (0.25% bromophenol blue, 0.25% xylene cyanol FF, 15% Ficol in water

Method

1. After harvesting the cells, samples are washed with phosphate-buffered saline (PBS) and pelleted by centrifugation.

2. The cell pellets are lysed in a solution containing 50 mM Tris–HCl (pH 8.0), 20 mM EDTA, 10 mM NaCl, 1% (w/v) sodium dodecylsulfate (SDS).

3. Lysates are incubated sequentially with 20 μg/ml RNase A at 37°C for 60 min and 100 μg/ml proteinase K, at 37°C for 3–5 h.

4. Lysates can then be applied directly on 1.2% agarose gel in TAE or TBE buffer in a horizontal gel support. In order to get clear DNA ladder images, it is important to optimize the cell number and the volume of cell lysis buffer because too many cells make the sample viscosity too high, and reduce the gel resolution.

5. Pour 1.2% agarose gel in TAE or TBE buffer in a horizontal gel support. Insert the comb and let the gel solidify.

6. Load samples to the gel wells, and perform electrophoresis.

7. Stain gels with 5 μg/ml ethidium bromide in TAE, TBE buffer, or water for 30 min. Resolved DNA fragments can be viewed on a UV transilluminator box and photographed.

8. Alternatively, DNA can be extracted from cell lysates by standard phenol/chloroform/isoamyl alcohol extraction procedures if sharp images cannot be obtained with whole-cell lysates.

5. Filter elution assay to measure apoptotic DNA fragmentation

DNA filter elution is commonly used to study the effects and mechanisms of action of chemotherapeutic drugs and carcinogens (38, 39). The basic DNA elution filter methods were originally designed to assay DNA damage in intact cells or tissues from living animals (38). More recently, DNA elution filter assays were adapted to study drug mechanisms in isolated nuclei (40, 41) and in a reconstituted cell-free system (13, 16). Altogether, the various elution methods are currently applied in the study of the DNA effects of a variety of anticancer agents (topoisomerase inhibitors, DNA cross-linking and alkylating drugs) in cells in culture, and in the analysis of DNA fragmentation associated with programmed cell death (apoptosis).

A variety of DNA lesions can be detected and quantitated by filter elution methods. Such lesions include DNA single-strand breaks (SSB), as well as double-strand breaks (DSB). SSB and DSB are either protein-associated (PASB) or protein-free ('frank') breaks. Alkaline elutions can also measure DNA–protein cross-links (DPC), interstrand DNA cross-links (ISC), and alkali-labile sites (ALS). More recently, we have used a simple filter elution method to quantitatively measure apoptosis-associated DNA fragmentation (42). We will only describe here this filter elution method. Details of the other alkaline elution methods can be found in detailed reviews and methods chapters (e.g. 38, 39, 43).

The filter elution assay can detect both internucleosomal and high molecular weight DNA fragmentation in apoptotic cells. Comparison with the pulse-field electrophoresis method indicated that filter elution can detect DNA fragmentation in the absence of internucleosomal DNA fragmentation and that DNA fragments up to 50–80 kb were detectable by filter elution.

5.1 Equipment

The basic equipment (38, 39) consists of filtration funnels (25 mm poly-ethylene filter holder, Swinnex, Millipore Corp.) connected and cemented with epoxy to a 50 ml polyethylene syringe (Swinnex funnel). The syringe orifice should be enlarged with a 0.125 in (3.2 mm) diameter drill to facilitate filling the upper section of the filter without trapping air. The bottom of the filter holder is connected to a 15 gauge stainless steel needle inserted through a rubber stopper to fit a filtration funnel holder (*Figure 1*). A set of elution funnels is mounted on a support, with a rack under this support to accommodate liquid scintillation vials (*Figure 1*).

5.2 Filters and solutions

- Filters: protein-adsorbent membrane filters 25 mm diameter, 0.8 μm pore size can be obtained from Gelman Sciences Inc. (Ann Arbor, MI) [PVC (polyvinyl chloride)/acrylic co-polymers, Metricel®] or from Poretics Corporation (Livermore, CA) (PVC).
- EDTA washing solution pH 10.0:
  (a) Prepare Na_2EDTA (0.1 M) from 37.2 g $Na_2EDTA.2H_2O$(ethylene-diaminetetraacetic acid disodium salt: dihydrate) and 8 g NaOH made up to 1 litre with H_2O. Adjust pH to 10.0 with NaOH (5 M).
  (b) For 500 ml of 0.02 M EDTA, mix 100 ml of Na_2EDTA (0.1 M) and 400 ml of H_2O and adjust the pH to 10.0.
- LS-10 (lysis solution pH 10.0): add 116.8 g NaCl, 400 ml 0.1 M Na_2EDTA,

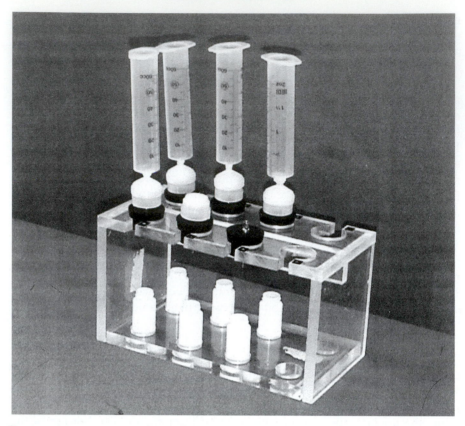

Figure 1. Photograph of a filter elution assay set-up, including a custom-designed rack to hold elution funnels and scinitillation vials. Front right: a rubber stopper with central needle (see text); front middle: the same with the assembled lower part of a Swinnex filter holder; front left and second row: four assembled elution funnels.

6.7 ml of 30% sarkosyl solution, and distilled water to a final volume of 1 litre. Adjust to pH 10.0 with NaOH/HCl.

- nucleus buffer: add 0.136 g KH_2PO_4, 1.016 g $MgCl_2$, 8.76 g NaCl, 0.380 g EGTA, and distilled water to a final volume of 1 litre. Adjust the pH to 6.4 with NaOH/HCl. Add DTT (dithiothreitol) to 0.1 mM final concentration immediately before use.
- nucleus buffer + Triton X-100: add 1 ml of Triton X-100 (30% stock solution) to 100 ml of nucleus buffer.

5.3 DNA labelling and preparation of experimental cell cultures

The DNA of exponentially growing cells is labelled with [^{14}C]thymidine (0.01–0.02 μCi/ml) for approximately one doubling time. Cells are then post-incubated for at least 4 h in isotope-free medium to chase radioactivity into high molecular weight DNA.

Control and drug-treated monolayer cells are detached by gentle scraping with a rubber 'policeman' and dispersed by repeated pipetting in their medium. Suspended cells are dispersed by repeated pipetting in their medium. Typically, $0.5–1.0 \times 10^6$ cells (corresponding to at least 5000 dpm.) are diluted in 5–10 ml of ice-cold Hank's balanced salt solution and loaded on to an elution funnel. The first 3 ml of cell suspension are loaded with a 1 ml automatic micropipette (the pipette tip needs to be inserted into the Swinnex hole but not too deep, to avoid puncturing the filter).

5.4 Protocols

Protocol 3. DNA fragmentation assay

Reagents

- Hank's balanced salt solution (HBSS)
- PBS
- LS-10 lysis solution
- 0.02 M EDTA solution

Method

1. Prepare Swinnex elution funnels and solutions (see Sections 5.1 and 5.2).
2. Load the cells or isolated nuclei (~0.5×10^6) directly on to the filter and wash immediately with an additional 5 ml of Hank's balanced salt solution (HBSS) or PBS. Allow this solution to flow through by gravity. This represents the 'supernatant' or 'soluble' fraction (S).
3. Add 5 ml LS-10 lysis solution and collect immediately. This represents the 'lysis' fraction (L).
4. Wash the lysis solution by adding 5 ml of 0.02 M EDTA solution, pH 10.0. Collect. This represents the 'wash' fraction (W).
5. Remove the filter and process the samples as indicated in Section 5.5. The 'filter' vial is labelled (F).

Protocol 4. Preparation of isolated nuclei

DNA filter elution assays can also be done using isolated nuclei. This application can be used to study nuclease activities and examine bio-chemical pathways that regulate nuclear apoptosis (see *Protocol 5*).

Reagents

- nucleus buffer
- 0.3% Triton X-100

Protocol 4. *Continued*

Method

1. Wash the cells twice in ice-cold nucleus buffer (see Section 5.2).

2. Incubate the cells ($\sim$1 $\times$ 10^6/ml) on ice with gentle rocking for 10 min in nucleus buffer containing 0.3% Triton X-100.

3. Centrifuge (450 *g*, 10 min at 4°C) to pellet the nuclei. Supernatants contain the cytoplasm and plasma membrane debris.

4. Wash the nuclear pellets twice by centrifugation/resuspension in lysis buffer without Triton X-100. Isolated nuclei are then ready to be used for experiments.

Protocol 5. Reconstituted cell-free system using isolated nuclei

Endonuclease activities that are associated with apoptosis can be detected from a reconstituted cell-free system (16, 37, 42).

Method

1. Prepare the nuclear and cytoplasmic fractions from control and apoptotic cells as in *Protocol 4*.

2. Centrifuge (12 000 *g*, 10 min at 4°C) the supernatants obtained as in *Protocol 4* to obtain soluble components of cytoplasmic fractions.

3. Incubate the control nuclei with either control or drug-treated (apoptotic) cytoplasmic fractions at 30°C for up to 30 min.

4. Load these extracts directly on to filters and carry out the DNA filter binding assay as in *Protocol 3* above.

5.5 Counting samples and computations

- All fractions contained in scintillation bottles are collected and the volume adjusted to $\sim$5 ml to keep the ratio of water to liquid scintillation cocktail at around 0.5 for optimum counting efficiency.

- Filters are placed at the bottom of a scintillation bottle and then add 0.4 ml of 0.1 M HCl is added. The vial is sealed and heated for 1 h at 65°C to depurinate the DNA. After removing the vials from the oven, 2.5 ml of 0.4 M NaOH is added and the vial is allowed to stand at room temperature for 45–60 min to release the DNA from the filter into solution and to allow maximum counting efficiency. Then, 2 ml H_2O is added to the 'filter' vial.

- Scintillation cocktail (10 ml) is added [Aquassure R (New England Nuclear)]. Radioactivity is then measured in each fraction by double-isotope counting (^{3}H for internal standard cells and ^{14}C for experimental cells) by liquid scintillation spectrometry.

- DNA fragmentation is determined as the fraction of labelled DNA in the lysis fraction + EDTA wash relative to total intracellular DNA. Results are expressed as the percentage of DNA fragmented in treated cells compared with DNA fragmented in untreated control cells (background) using the formula:

$$\%\text{DNA fragmentations} = [(F{-}F_o)/(1{-}F_o)] \times 100$$

where F and F_o represent DNA fragmentation in treated and control cells, respectively. Experimental data are usually plotted as the percentage of DNA fragmented versus the time, as shown in *Figure 2*.

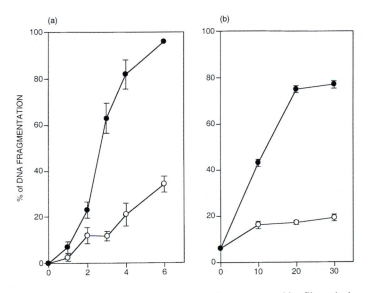

Figure 2. Apoptosis-associated DNA fragmentation measured by filter elution assay from cultured cells or from a reconstituted cell-free system. (a) HL60 cells were treated at 0.1 μM (open circles) and 1 μM (closed circles) staurosporine concentrations (14). DNA fragmentation was determined by the DNA filter elution assay at specified times after drug removal. (b) DNA fragmentation induced in a reconstituted cell-free system by cytoplasmic fractions obtained from HL60 cells treated with staurosporine (1 μM for 3 h). Cytoplasm from untreated HL60 cells (open circles) or staurosporine-treated cells (closed circles) was incubated with nuclei isolated from untreated cells at 30°C for the indicated times. DNA fragmentation was determined as described in Section 5.5.

References

1. Wyllie, A. H., Kerr, J. F. R., and Currie, A. R. (1980). *Int. Rev. Cytol.* **68**, 251–306.
2. Wyllie, A. H. (1995). *Curr. Opin. Genet. Dev.* **5**, 97–104.
3. Wyllie, A. H. (1980). *Nature* **284**, 555–6.

4. Arends, M. J., Morris, R. G., and Wyllie, A. H. (1990). *Am. J. Pathol.* **136**, 593–608.
5. Falcieri, E., Martelli, A. M., Bareggi, R., Cataldi, A., and Cocco, L. (1993). *Biochem. Biophys. Res. Commun.* **193**, 19–25.
6. Catchpoole, D. R. and Stewart, B. W. (1993). *Cancer Res.* **53**, 4287–96.
7. Cohen, G. M., Sun, X. M., Snowden, R. T., Dinsdale, D., and Skilleter, D. N. (1992). *Biochem. J.* **286**, 331–4.
8. Oberhammer, F., Wilson, J. W., Dive, C., Morris, I. D., Hickman, J. A., Wakeling, A. E., Walker, P. R., and Sikorska, M. (1993). *EMBO J.* **12**, 3679–84.
9. Walker, P. R., Smith, C., Youdale, T., Leblanc, J., Whitfield, J. F., and Sikorska, M. (1991). *Cancer Res.* **51**, 1078–85.
10. Pandey, S., Walker, P. R., and Sikorska, M. (1994). *Biochem. Cell Biol.* **72**, 625–9.
11. Cohen, G. M., Sun, X. M., Fearnhead, H., MacFarlane, M., Brown, D. G., Snowden, R. T., and Dinsdale, D. (1994). *J. Immunol.* **153**, 507–16.
12. Walker, P. R. and Sikorska, M. (1997). *Biochem. Cell Biol.* **75**, 287–99.
13. Bertrand, R., Sarang, M., Jenkins, J., Kerrigan, D., and Pommier, Y. (1991). *Cancer Res.* **51**, 6280–5.
14. Bertrand, R., Solary, E., O'Connor, P., Kohn, K. W., and Pommier, Y. (1994). *Exp. Cell Res.* **211**, 314–21.
15. Solary, E., Bertrand, R., and Pommier, Y. (1994). *Leuk. Lymph.* **15**, 21–32.
16. Solary, E., Bertrand, R., Kohn, K. W., and Pommier, Y. (1993). *Blood* **81**, 1359–68.
17. Pommier, Y., Leteurtre, F., Fesen, M. R., Fujimori, A., Bertrand, R., Solary, E., Kohlhagen, G., and Kohn, K. W. (1994). *Cancer Invest.* **12**, 530–42.
18. Gavrieli, Y., Sherman, Y., and Ben-Sasson, S. A. (1992). *J. Cell Biol.* **119**, 493–501.
19. Peitsch, M. C., Muller, C., and Tschopp, J. (1993). *Nucl. Acids Res.* **21**, 4206–9.
20. Walker, P. R., Leblanc, J., and Sikorska, M. (1997). *Cell Death Differ.* **4**, 506–15.
21. Peitsch, M. C., Polzar, B., Stephan, H., Crompton, T., MacDonald, H. R., Mannherz, H. G., and Tschopp, J. (1993). *EMBO J.* **12**, 371–7.
22. Rauch, F., Polzar, B., Stephan, H., Zanotti, S., Paddenberg, R., and Mannherz, H. G. (1997). *J. Cell Biol.* **137**, 909–23.
23. Gaido, M. L. and Cidlowski, J. A. (1991). *J. Biol. Chem.* **266**, 18580–5.
24. Montague, J. W., Gaido, M. L., Frye, C., and Cidlowski, J. A. (1994). *J. Biol. Chem.* **269**, 18877–80.
25. Liu, Q. Y., Pandey, S., Singh, R. K., Ribecco, M., Borowy-Borowski, H., Smith, B., LeBlanc, J., Walker, P. R., and Sikorska, M. (1998). *Biochemistry* **37**, 10134–43.
26. Shiokawa, D., Ohyama, H., Yamada, T., and Tanuma, S. (1997). *Biochem. J.* **326**, 675–81.
27. Barry, A. and Eastman, A. (1993). *Arch. Biochem. Biophys.* **300**, 440–50.
28. Krieser, R. J. and Eastman, A. (1998). *J. Biol. Chem.* **273**, 30909–14.
29. Ribeiro, J. M. and Carson, D. A. (1993). *Biochemistry* **32**, 9129–36.
30. Khodarev, N. N. and Ashwell, J. D. (1996). *J. Immunol.* **156**, 922–31.
31. Pandey, S., Walker, P. R., and Sikorska, M. (1997). *Biochemistry* **36**, 711–20.
32. Bernardi, G. (1968). *Adv. Enzymol.* **31**, 1–7.
33. Enari, M., Sakahira, H., Yokoyama, H., Okawa, K., Iwamatsu, A., and Nagata, S. (1998). *Nature* **391**, 43–50.
34. Sakahira, H., Enari, M., and Nagata, S. (1998). *Nature* **391**, 96–9.

35. Liu, X., Li, P., Widlak, P., Zou, H., Luo, X., Garrard, W. T., and Wang, X. (1998). *Proc. Natl. Acad. Sci. USA* **95**, 8461–6.
36. Zhang, J., Liu, X., Scherer, D. C., van Kaer, L., Wang, X., and Xu, M. (1998). *Proc. Natl. Acad. Sci. USA* **95**, 12480–5.
37. Yoshida, A., Pourquier, P., and Pommier, Y. (1998). *Cancer Res.* **58**, 2576–82.
38. Kohn, K. W., Ewig, R. A. G., Erickson, L. C., and Zwelling, L. A. (1981). In *DNA repair: a laboratory manual of research procedures* (ed. E. C. Friedberg and P. C. Hanawalt), pp. 379–401. Marcel Dekker Inc., New York.
39. Bertrand, R. and Pommier, Y. (1995). In *Cell growth and apoptosis: a practical approach* (ed. G. Studzinski), pp. 96–117. IRL Press, Oxford University Press, Oxford.
40. Filipski, J., Yin, J., and Kohn, K. W. (1983). *Biochim. Biophys. Acta* **741**, 116–22.
41. Pommier, Y., Schwartz, R. E., Kohn, K. W., and Zwelling, L. A. (1984). *Biochemistry* **23**, 3194–201.
42. Bertrand, R., Kohn, K. W., Solary, E., and Pommier, Y. (1995). *Drug Develop. Res.* **34**, 138–44.
43. Kohn, K. W. (1996). *Bioessays* **18**, 505–13.

4

Analysis of cell death by flow and laser-scanning cytometry

ZBIGNIEW DARZYNKIEWICZ, ELZBIETA BEDNER,
and XUN LI

1. Introduction

Flow cytometry, by providing the possibility of rapid, accurate, and unbiased measurements of a variety of cell components on a cell by cell basis, has become an indispensable methodology in the analysis of cell death (see reviews in refs 1–6). One area of application of flow cytometry is in studies of the mechanism of cell death. In this application, flow cytometry is primarily used to immunocytochemically detect and measure the cellular levels of proteins such as members of the Bcl-2 family, proto-oncogenes c-*myc* or *ras*, tumour suppressor genes p53, pRB, and other molecules that play a role in cell death. It is also used in studies of cell function, particularly mitochondrial metabolism, which is closely associated with mechanisms regulating cell sensitivity to apoptosis (e.g. 7, 8). The major virtue of flow cytometry in this application is that it offers the possibility of multiparametric analysis of a multitude of cell attributes. This allows one to study the mutual relationship between the measured constituents. When one of the measured attributes is cellular DNA content, a relationship of other parameters to the cell cycle position or DNA ploidy is analysed. Because individual cells are measured the intercellular variability can be assessed, cell subpopulations identified, and rare cells detected.

Another area of application of flow cytometry is in the identification and quantitation of apoptotic or necrotic cells. Their recognition is generally based on the presence of a particular biochemical or molecular marker that is characteristic for either apoptosis or necrosis. A variety of methods have been developed, especially for identification of apoptotic cells. The apoptosis-associated changes in cell size and granularity can be detected by analysis of laser light scattered by the cell in forward and side directions (9). Some of the methods rely on the apoptosis-associated changes in the distribution of plasma membrane phospholipids (5, 10). Others measure the transport function of the plasma membrane. Still other methods probe the mitochondrial transmembrane potential, which decreases early during apoptosis (7, 8, 11). Endonucleolytic DNA degradation that results in extraction of low MW DNA from the cell provides yet another marker of apoptosis. Apoptotic cells are then recognized either by their fractional DNA content (12, 13) or by the presence of DNA strand breaks which can be detected by labelling their 3'-OH ends with fluorochrome-conjugated nucleotides in a reaction utilizing exogenous terminal deoxynucleotidyl transferase (TdT) (14–18).

The common drawback of flow cytometric methods is that the identification of apoptotic or necrotic cells relies on a single parameter reflecting a change in a biochemical or molecular feature of the cell that is assumed to represent apoptosis or necrosis. However, such a feature may be absent when apoptosis is atypical, as is known to occur, e.g. in cells of epithelial and fibroblast lineage (19–21). Atypical apoptosis is also caused by agents that inhibit apoptotic effectors. For example, induction of apoptosis by an inhibitor of the endo-nuclease results in a lack of DNA fragmentation, while inhibitors of proteases prevent degradation of particular proteins such as 'death substrates' (e.g. ref. 22). Identification of apoptotic cells based on the missing attribute(s) (e.g. DNA fragmentation or degradation of a particular protein) is impossible in such cases. Therefore, the characteristic changes in cell morphology, as originally described (23, 24) and discussed in Chapter 2, still remain the gold standard for recognition of apoptotic cell death.

The laser-scanning cytometer (CompuCyte, Cambridge, MA) is a microscope-based cytofluorometer that offers advantages of both flow cyto-metry and image analysis (for reviews see refs 25 and 26). Thus, fluorescence of individual cells is measured rapidly by laser-scanning cytometry (LSC) and with an accuracy similar to that of flow cytometry. Since cell position on the slide is recorded together with other measured cell parameters in a list-mode fashion, the cells can be relocated after measurements and re-examined visually or subjected to image analysis (e.g. 27, 28). Furthermore, the geometry of cells attached to the slide, especially when flattened by cytocen-trifugation, is more favourable for their morphometric analysis than when in suspension. More information on cell morphology, therefore, can be obtained by LSC than by flow cytometry. In the analysis of apoptosis and necrosis, an opportunity to examine measured cells visually, as offered by LSC, is of par-ticular value. Furthermore, cell analysis on slides eliminates cell loss, which generally occurs during repeated centrifugations in sample preparation for flow cytometry.

Several flow cytometric methods developed for the identification of apoptotic and necrotic cells have been modified and adapted so that they may be used for LSC (e.g. 29–31). These modifications and changes in methodology re-quired by the adaptation of flow cytometric methods to LSC are presented in this chapter. Also discussed are difficulties in the measurement of apoptosis or necrosis, as well as common errors in the analysis and interpretation of data.

2. Preparation of cells for analysis by LSC. Detection of apoptotic cells based on changes in nuclear chromatin

Several assays of apoptosis by LSC require cell fixation. In these assays the cells are attached to microscope slides by standard methods that include smear films, tissue sections, 'touch' preparations from the tissues, or cyto-centrifuging cell suspensions. Cytocentrifugation is often preferred over 'touch' or smear preparations because it flattens the cells on the slides, so more morphological details are revealed.

LSC measurement of total nuclear or cellular fluorescence is done by integration of the light intensity of individual pixels over the area of nucleus and/or cytoplasm (25, 26). Thus, the intensity of individual pixels, as well as the fluorescence area (number of pixels), is being measured. The intensity of the maximal pixel within the measured area is also recorded. Because of the high degree of chromatin condensation in apoptotic cells, DNA stains with a greater intensity per unit of the projected nuclear area in these cells (hyper-chromasia). The maximal pixel value of the DNA-associated fluorescence measured in the chromatin of apoptotic cells, therefore, is greater than in the non-apoptotic nuclei (30). The situation is analogous to that of the chromatin of mitotic cells, which is also strongly condensed. Apoptotic cells, therefore, similarly to mitotic cells (31), can be identified by high values of the maximal pixel of DNA-associated fluorescence (*Figure 1*).

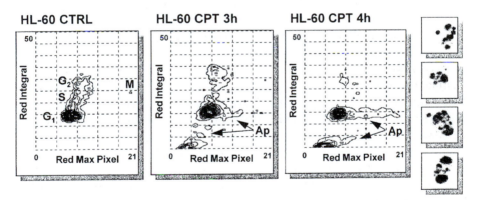

Figure 1. Detection of apoptotic cells by LSC based on DNA hyperchromicity in condensed chromatin. Apoptosis was induced by treatment of HL60 cells with 0.15 μM camptothecin (CPT) for 3 or 4 h, as described (16, 17). The cells were stained with PI according to *Protocol 1*. In the untreated, control culture (CTRL) only mitotic cells (M) have a high value of red fluorescence maximal pixel. In the CPT-treated cultures, concomitant with a loss of S phase cells, the cells with high values of red maximal pixel become apparent. After relocation, when viewed by fluorescence microcopy, their morphology, as well as the morphology of the cells with fractional DNA content, was typical of apoptotic cells (shown in the four illustrations on the right).

Protocol 1. Identification of apoptotic cells by changes in nuclear chromatin

Reagents
- stock solution of propidium iodide (PI): dissolve 1 mg of PI (available from Molecular Probes, Inc., Eugene, OR) in 1 ml of distilled water. This solution can be stored at 0–4°C for weeks.
- prepare 1% formaldehyde solution in PBS. This solution should also be made fresh.

- prepare a working solution of PI by adding 200 μl of the stock solution of PI into 10 ml of PBS. Add, and dissolve in this solution, 2 mg RNase A (the final concentration of PI and RNase should be 20 μg/ml and 0.2 mg/ml, respectively). This solution should be made fresh.

Method

1. Deposit the cells on the microscope slide, preferably by cytocentrifugation. To attach cells by cytocentrifugation add 300 μl of cell suspension in tissue culture medium (with serum) containing approximately 20 000 cells into a cytospin (e.g. Shandon Scientific, Pittsburgh, PA) chamber. Cytocentrifuge at 1000 r.p.m. (~110g) for 6 min.

2. Without allowing the cytospun cells to dry completely, fix them by immersing the slides in a Coplin jar containing 1% formaldehyde in PBS, on ice for 15 min.

3. Transfer the slides into Coplin jars containing 70% ethanol. The cells may be stored in ethanol for several days.

4. After fixation, rinse the slides in PBS for 5 min. Stain the cells with the working solution of PI for 20 min at room temperature. Mount the cells under a coverslip in the working solution of PI and seal the preparation with melted paraffin or nail polish. Measure the cellular red fluorescence (>600 nm; integrated fluorescence and maximal pixel) by LSC, illuminating the cells at 488 nm.

Identification of apoptotic cells by LSC based on the highest pixel value for DNA-associated fluorescence, as shown in *Figure 1*, is done in conjunction with DNA content analysis. Thus, DNA ploidy and/or the cell cycle position of apoptotic and non-apoptotic cells, can be determined at the same time as the estimate of the apoptotic index. Furthermore, the staining procedure is simple and can be combined with analysis of other constituents of the cell when the latter are probed with fluorochromes of another colour. The disadvantage of this approach is that it cannot discriminate between mitotic and apoptotic cells. Also, early G1 (post-mitotic) cells may have high fluorescence intensity of the maximal pixel (31). The distinction between apoptotic and mitotic cells is critical during treatment with agents such as taxol or other mitotic blockers, i.e. when mitotic cells undergo apoptosis. However, visual examination of the cells, or analysis of other morphometric features, such as nuclear to cytoplasm ratio, nuclear or cell area, forward light scatter, etc., as offered by LSC, can be helpful in these instances.

Several cytometric methods designed to identify apoptotic cells, or to study molecular or metabolic events associated with apoptosis, probe the cells that have vital functions preserved and therefore cannot be fixed prior to analysis. Analysis of plasma membrane transport function (e.g. 9, 32), detection of phosphatidylserine on the plasma membrane (5, 10; see Chapter 7), or probing the mitochondrial metabolism (7, 8; see Chapter 8) all require unfixed cells. Suspensions of live cells in appropriately prepared reaction media are generally used when such analyses are done by flow cytometry. In the case of LSC, however, the measured cells have to be attached to a microscope slide to be relocated for morphological examination, additionally probed by another fluorochrome(s), or stained with a light-absorbing dye for analysis by light microscopy.

Two different approaches can be used to attach live cells to microscope slides. The first involves direct cell culturing on microscope slides or coverslips. Culture vessels are commercially available that have a microscope slide for the bottom of the chamber (e.g. 'Chamberslide', Nunc, Inc., Naperville, Il). The cells growing in these chambers spread and attach to the slide surface. Chambers made on glass rather than plastic slides are preferred as the latter often have high autofluorescence. Alternatively, the cells can be grown on coverslips, which then can be inverted over shallow wells on the microscope slides for analysis by LSC. This mode of attaching cells to slides is, of course, applicable only to cells that normally grow on the surface of flasks, e.g. cells of epithelial or fibroblast lineage, macrophages/monocytes, etc. It should be stressed, however, that because the cells detach during late stages of apoptosis, many apoptotic cells may be selectively lost if the analysis is limited to the attached cells only.

Cells that grow in suspension can be attached to glass slides by electrostatic forces. This is done by incubating them on microscope slides in the absence of any serum or serum proteins (27–29, 33). A short (15–20 min) incubation of the cell suspension in PBS, at room temperature and 100% humidity, in shallow wells on the microscope slides, is generally adequate to attach the cells to the slide surface. The live cells thus attached can be subjected to surface immmunophenotyping (33), viability tests, or intracellular enzyme kinetics assays (27–29). Following fixation, e.g. in formaldehyde, the electrostatically attached cells become firmly attached by the fixative. When such preparations are restained, a large majority of the cells (>95%) is found to be still attached and can be relocated by LSC (33). However, as in the case of cell growth on glass, the late apoptotic cells have a tendency not to attach, or may detach after the initial attachment.

3. Light-scattering properties of apoptotic and necrotic cells

Intersection of a cell with the light of the laser beam in a flow cytometer leads to light scatter, and analysis of the signal of the scattered light reveals information about cell size and structure (34). The possibility of the analysis of light scattered in the forward and at right angles ('side scatter'; 90°) directions is a built-in feature of every commercially available flow cytometer utilizing laser illumination, and is a routine measurement, either alone or in conjunction with the measurement of cell fluorescence. Forward light scatter correlates well with cell size, whereas side scatter bears information on the cell light refractive and reflective properties and reveals optical inhomogeneity of the cell structure, such as that resulting from condensation of the cytoplasm or nucleus, granularity, etc. Side scatter, however, cannot be measured by LSC.

As a consequence of cell shrinkage, a decrease in forward light scatter is observed at a relatively early stage of apoptosis (*Figure 2*). Initially, there is little change in side scatter during apoptosis, though in some cell systems an increase in intensity of the side scatter signal is seen, reflecting, perhaps, chromatin and cytoplasm condensation and nuclear fragmentation. When apoptosis is more advanced and the cells become small, the intensity of the side scatter decreases, just as the forward scatter. Late apoptotic cells are thus characterized by a markedly diminished intensity for both the forward and

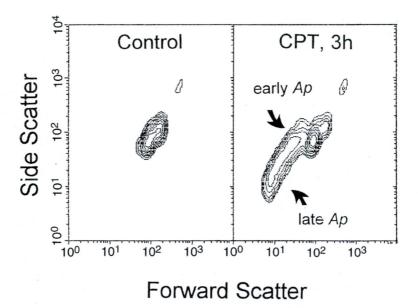

Figure 2. Light scattering properties of apoptotic cells. To induce apoptosis HL60 cells were treated with CPT for 3 h (16, 17). Their light scattering properties were measured by FACScan (Becton Dickinson, San Jose, CA). The heterogeneity of apoptotic cells is reflected by the large variability in the scattered light. The early apoptotic cells (Ap) have diminished forward scatter but show no change in side scatter.The cells that are more advanced in apoptosis have both the forward and side scatter greatly decreased.

side scatter signals. In contrast to apoptosis, cell swelling, which occurs early during cell necrosis, is detected by a transient increase in forward light scatter. Rupture of the plasma membrane and leakage of the cytosol during subsequent steps of necrosis correlates with a marked decrease in the intensity of both forward and side scatter signals.

Because there is a great variability in the light scattering properties of individual cells, an exponential scale (logarithmic amplifiers) should be used during scatter measurements. Analysis of light scatter is often combined with other assays, most frequently with surface immunofluorescence (e.g. to identify the phenotype of the dying cell), or with another marker of apoptosis. It should be mentioned, however, that the light scatter changes alone are not a very specific marker of apoptosis or necrosis. Mechanically broken cells, isolated nuclei, cell debris, and individual apoptotic bodies may also have diminished light scattering properties. Therefore, the analysis of light scatter should be combined with measurements that can provide a more definite identification of apoptotic or necrotic cells.

4. Analysis of mitochondrial transmembrane potential ($\Delta\psi_m$)

Decrease in $\Delta\psi_m$ is one of the early events of apoptosis (7, 8, 11, see also Chapter 8). Several types of membrane-permeable lipophilic cationic fluorochromes can serve as probes of $\Delta\psi_m$ in flow or laser-scanning cytometry. When live cells are incubated in their presence the probes accumulate in mitochondria and the extent of their uptake, measured by the intensity of the cellular fluorescence, is considered to reflect $\Delta\psi_m$. Rhodamine 123 (Rh123) and 3,3'-dihexiloxa-dicarbocyanine [DiOC$_6$ (3)] are the fluorochromes most frequently used thus far to probe $\Delta\psi_m$ (7, 8, 11, 35). A combination of Rh123 and propidium iodide (PI) was originally proposed as a cell viability assay discriminating between live cells that stain with Rh123 only (green fluorescence), dead or dying cells whose plasma membrane integrity was compromised (cells with damaged plasma membrane, late apoptotic, and necrotic cells) that stain only with PI (red fluorescence), and early apoptotic cells that show somewhat increased stainability with PI but also stain with Rh123 (35). The specificity of Rh123 and DiOC$_6$ (3) as mitochondrial probes is greater when these fluorochromes are used at low concentration. Still another probe of $\Delta\psi_m$ is the J-aggregate-forming lipophilic cationic fluorochrome 5,5',6,6'-tetrachloro-1,1',3,3'-tetrathylbenzimidazolcarbocyanine iodide (JC-1) (7). Its binding to mitochondria is detected by the shift in colour of fluorescence from green, which is characteristic of its monomeric form, to orange, which reflects its aggregation in mitochondria, driven by the transmembrane potential (7).

Protocol 2. Measurement of mitochondrial transmembrane potential

Reagents and stock solutions

- 3,3'-dihexiloxa-dicarbocyanine [DiOC$_6$ (3)], rhodamine 123 and/or 5,5',6,6'-tetrachloro-1,1',3,3'-tetraethylbenzimidazolcarbocyanine iodide (JC-1) (Molecular Probes)
- PI (Molecular Probes)
- stock solution of Rh123: dissolve 0.1 mg of Rh123 in 1 ml distilled water
- stock solution of DiOC$_6$ (3): dissolve 5 μg of DiOC$_6$ (3) in 1.0 ml ethanol

- stock solution of JC-1: dissolve 1 mg JC-1 in 1 ml DMSO
- stock solution of PI: dissolve 1 mg PI in 1 ml distilled water

Each of the above stock solutions is stable and can be stored at 0–4°C in the dark for weeks.

- working solution of DiO$_6$ (3): add 5.0 μl of the stock solution of DiO$_6$ (3) and 5 μl of the stock solution of PI into 1 ml PBS

Methods

A. *Staining with Rh123 and PI and analysis by flow cytometry*

1. Add 1.0 μl of Rh123 stock solution to approximately 10^6 cells suspended in 1 ml of tissue culture medium (or HBSS) and incubate for 10 min at 37°C in the dark.

2. Add 20 μl of the PI stock solution and keep for 5 min at room temperature in the dark.

3. Analyse cells by flow cytometry.

   - use excitation in blue light (e.g. 488 nm line of the argon ion laser)
   - use light scattering to trigger cell measurement
   - measure green (Rh123) fluorescence at 530±20 nm
   - measure red (PI) fluorescence at >600 nm

Analysis by LSC

To be analysed by LSC the cells have to be first attached to a microscope slide, then incubated with Rh123 and PI. To attach the cells:

1. Rinse the cells to be free of serum (or bovine serum albumin) and resuspend them in PBS to have 2×10^5–10^6 cells/ml.

2. Take 50 μl of this suspension and deposit it within a shallow well on the microscope slide. The slides with wells can be made by preparing a strip of Parafilm 'M' (American National Can, Greenwich, CT) of the size of the slide, cutting a hole 2.0 cm × 0.5 cm in the middle of this strip, placing the strip on the microscope slide and heating the slide on a warm plate until the Parafilm starts to melt. Alternatively, the well can be made by drawing a border with a hydrophobic marker ('Isolator', Shandon Scientific, Pittsburgh, PA). The cell suspension is placed within the well on a slide and is maintained at room temperature at 100% humidity for 10 min to allow the cells to attach electrostatically to the slide surface.

3. The cells thus attached are then incubated for 10 min with Rh123 and then with PI as described above for cells in suspension. A coverslip is then placed over the well and using illumination at 488 nm the cells' green and red fluorescence is measured by LSC.

4. If desired, the cells, while still attached to the slide after the measurement, can be fixed (e.g. in formaldehyde or ethanol) for subsequent staining with other dye(s) and/or visual inspection or measurement, following their relocation ('merge') on the slide.

B. *Staining with DiOC$_6$ (3) and PI and analysis by flow cytometry*

1. Suspend the cell pellet ($\sim$10^6 cells) in a working solution of DiOC$_6$ (3).

2. Incubate for 15 min at room temperature in the dark.

3. Measure the cells' green and red fluorescence as described above for Rh123 and PI.

Analysis by LSC

Follow the procedure as described above for Rh123 and PI, except that after attaching the cells to the microscope slide incubate them with the working solution of DiOC$_6$ (3) for 15 min.

C. *Staining with JC-1 and analysis by flow cytometry*

1. Suspend the cell pellet ($\sim$10^6 cells) in 1 ml of tissue culture medium with 10% serum.

2. Add 10 μl of the stock solution of JC-1. Vortex cells intensely during addition of JC-1 and for the next 20 sec.

3. Incubate cells for 10 min at room temperature in the dark.

4. Measure the cells' fluorescence by flow cytometry.
   - excite fluorescence with blue light (488 nm line of argon laser)
   - measure green fluorescence at 530$\pm$20 nm
   - measure orange fluorescence at 570$\pm$20 or above (long pass filter) 570 nm

Analysis by LSC

1. Prepare the cells and incubate them with JC-1 in suspension as described above in steps 1–3.

2. Place 50 μl of this suspension into a well on the microscope slide. Cover with a coverslip.

3. Analyse cells by LSC.
   - use excitation in blue light (488 nm)
   - use forward scattering as a contouring (triggering) parameter
   - measure green fluorescence at 530$\pm$20 nm
   - measure orange fluorescence at 570$\pm$20 or above (long pass filter) 570 nm

The decrease in $\Delta\psi_m$ which occurs during apoptosis can be measured by different markers, as discussed above, and shown in *Figure 3*. A combination of PI and $DiOC_6$ (3) identifies non-apoptotic cells that stain only green, early apoptotic cells whose green fluorescence is markedly diminished, and late apoptotic or necrotic cells that stain with PI showing red fluorescence. Likewise, a combination of Rh123 and PI labels live non-apoptotic cells green, early apoptotic cells dim green, and late apoptotic and necrotic cells red. The change in binding of JC-1 is manifested by a loss of orange fluorescence, which represents the aggregate binding of this dye in mitochondria. Green fluorescence of JC-1 also decreases during apoptosis, although to a lesser degree than orange. Because still another mitochondrial probe, nonyl acridine orange, reports mitochondrial mass but is not sensitive to $\Delta\psi_m$ (7) it is possible

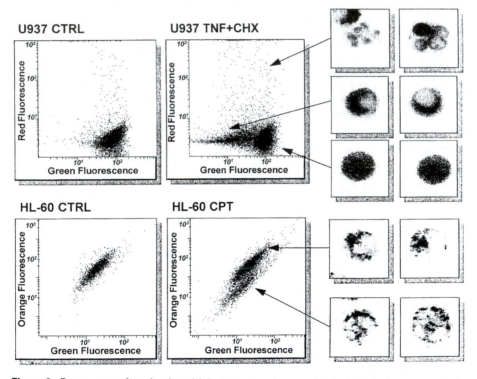

Figure 3. Decrease of mitochondrial transmembrane potential during apoptosis measured by LSC. Apoptosis of U-937 cells was induced by their incubation with 5 ng/ml TNF-α combined with with 5 μg/ml of cycloheximide for 3 h; the cells were then incubated with $DiOC_6$ (3) and PI as described in *Protocol 2*. HL60 cells were treated with 0.15 μM CPT for 4 h to induce apoptosis and then stained with JC-1 as described in *Protocol 2*. The drop in mitochondrial transmembrane potential ($\Delta\psi_m$), which characterizes apoptotic cells, is reflected by a decrease in intensity of the green fluorescence of $DiOC_6$ (3)-stained cells (CHX, top right panel) and loss of orange fluorescence of JC-1 stained cells (bottom right panel). Note that very few cells stain with PI (red fluorescence).

to analyse both the mitochondrial mass and $\Delta\psi_m$ in the same cells. Although apoptotic cells are sometimes more flattened on the slides and may have a greater diameter, their forward light scatter is diminished compared with the non-apoptotic cells.

5. Detection of apoptotic cells exposing phosphatidylserine on plasma membrane

In live, non-apoptotic cells phospholipids of the plasma membrane are asymmetrically distributed between the inner and outer leaflets of the membrane. Thus, while phosphatidylcholine and sphingomyelin are exposed on the external leaflet of the lipid bilayer, phosphatidylserine is almost exclusively located on the inner surface. Early during apoptosis this asymmetry is broken and phosphatidylserine becomes exposed on the outside of the plasma membrane (5, 10, see also Chapter 7). Because the anticoagulant protein annexin V binds with high affinity to negatively charged phospholipids like phosphatidylserine, the fluoresceinated annexin V has found an application as a marker of apoptotic cells, in particular for their detection by flow cytometry (5). In the course of apoptosis the cells become reactive with annexin V after the onset of chromatin condensation but prior to the loss of plasma membrane ability to exclude cationic dyes such as PI. Therefore, by staining cells with a combination of fluorescein-conjugated annexin V and PI, it is possible to detect unaffected, non-apoptotic cells (annexin V negative/PI negative), early apoptotic cells (annexin V positive/PI negative), and late apoptotic or necrotic cells (PI positive) by flow or laser-scanning cytometry.

Protocol 3.   Measurement of changes in plasma membrane

Reagents

- dissolve fluorescein-conjugated annexin V [1:1 stoichiometric complex, available from BRAND Applications, The Netherlands (cat. no. A-700)] in binding buffer [10 mM Hepes (*N*-2-hydroxyethylpiperazine-*N*-2-ethane-sulfonic acid)–NaOH, pH. 7.4, 140 mM NaCl, 2.5 mM CaCl$_2$] at concentration of 1.0 μg/ml This solution has to be prepared fresh prior to use.

- PI stock solution: dissolve 1 mg PI in 1 ml distilled water. The solution is stable for months when stored in the dark at 0–4 °C.

Method

1. Suspend 10^5–10^6 cells in 1 ml of fluorescein-conjugated annexin V in binding buffer for 5 min.

2. Add PI solution to the cell suspension prior to analysis to have a final PI concentration of 1.0 μg/ml.

Protocol 3. *Continued*

3. Analyse cells by flow cytometry.
   - use excitation in blue light (e.g. 488 nm line of the argon ion laser)
   - light scattering (forward vs. side) may be used to trigger cell measurements
   - measure green fluorescein–annexin V fluorescence at 530±20 nm
   - measure red (PI) fluorescence at >600 nm

4. Cell analysis by LSC. To be analysed by LSC the cells should be stained in suspension, as described above, then placed on a microscopic slide under a coverslip for fluorescence measurement. Alternatively, the cells should be initially attached electrostatically to a microscope slide, as described above (see Section 2) and then subjected to staining with fluoresceinated annexin V and PI. It is critical that cells should not dry or be mechanically damaged during the attachment and staining procedure.

Live, non-apoptotic cells have low green (fluorescein) fluorescence and little or no red PI fluorescence. Cells undergoing apoptosis stain green but continue to be negative for PI staining; such cells often have low forward and high side scatter (*Figure 4*). Isolated nuclei, or cells with damaged membranes should stain rapidly and brightly with PI. Note that cells with damaged membranes may also be positive for annexin V since phosphatidylserine on the inside of the plasma membrane is available to the ligand. Thus, damaged cells may be PI positive/annexin V negative or PI positive/annexin V positive. Any procedure which affects the integrity of the membrane will allow access of annexin V to the inner leaflet of plasma membrane, resulting in cells which will be scored as positive for apoptosis. The use of proteolytic enzymes to disrupt cell clumps or to remove adherent cells from culture walls was reported to affect the binding of fluoresceinated annexin V. As a result, care should be exercised in the use of such preparatory procedures. Longer incubation times will also allow cells which still maintain membrane integrity to become positive for PI since this dye will enter intact cells, although very slowly. Therefore, if flow cytometric analysis is not performed quickly after staining, do not add PI until 5 min before measurement. Apoptosis, itself, is an ongoing process. As a result, cells stained with fluorescein–annexin V should not be kept for prolonged lengths of time (>1 h) prior to measurement.

6. Detection of apoptotic cells based on their fractional DNA content

Activation of an endonuclease causes DNA fragmentation (see Chapter 3) and such DNA can be extracted from the cells following their fixation and permeabilization such that less DNA in apoptotic cells stains with any DNA

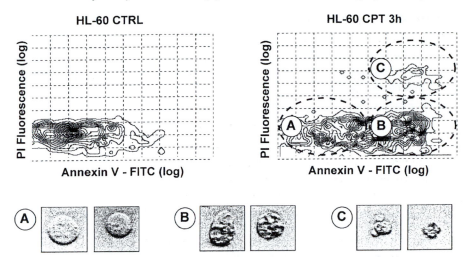

Figure 4. The distinction between live, early apoptotic, and late apoptotic/necrotic cells by LSC after their staining with fluoresceinated annexin V and PI as described in *Protocol 3*. Apoptosis of HL60 cells was induced by their incubation with 0.15 μM of CPT for 3 h as described (16, 17). (A)The live, non-apoptotic cells neither stain with annexin V–FITC nor with PI. (B) Early apoptotic cells stain with fluorescein but exclude PI while (C) late apoptotic and necrotic cells stain with both dyes. The representative cells from the sectors A, B, or C of the contour map were relocated, viewed under blue light incident illumination, and their colour pictures were converted to grey scale.

fluorochrome (12, 13). The degree of DNA degradation varies depending on the stage of apoptosis, the cell type, and often the nature of the apoptosis-inducing agent. Hence, the extractability of DNA during the staining procedure also varies. It has been noted that a high molarity phosphate–citrate buffer enhances extraction of the fragmented DNA (36). This approach can be used to control the extent of DNA extraction from apoptotic cells to the desired level and to obtain their optimal separation by flow cytometry, as described in *Protocol 4*.

Protocol 4. Detection of fractional DNA content

Reagents
- DNA staining solution: dissolve 200 μg of PI in 10 ml of PBS and add 2 mg of DNase-free RNase A (boil RNase for 5 min if it is not DNase free). Prepare fresh staining solution before each use
- DNA extraction buffer: mix 192 ml of 0.2 M Na_2HPO_4 with 8 ml of 0.1 M citric acid; pH 7.8

Method

1. Fix the cells in suspension in 70% ethanol by adding 1 ml of cells suspended in PBS (1–5×10^6 cells) into 9 ml of 70% ethanol in a tube on ice. Cells can be stored in fixative at –20°C for several weeks.

Protocol 4. *Continued*

2. Centrifuge cells (200 g, 3 min), decant ethanol, suspend cells in 10 ml of PBS, and centrifuge (300 g, 5 min).

3. Suspend cells in 0.5 ml PBS, into which you may add 0.2–1.0 ml of the DNA extraction buffer.

4. Incubate at room temperature for 5 min, centrifuge.

5. Suspend the cell pellet in 1 ml of DNA staining solution.

6. Incubate the cells for 30 min at room temperature.

7. Analyse the cells by flow cytometry.
   - use 488 nm laser line (or a mercury arc lamp with a BG12 filter) for excitation
   - measure red fluorescence (>600 nm) and forward light scatter

Alternative methods

Cellular DNA may be stained with other fluorochromes instead of PI, and other cell constituents may be counterstained in addition to DNA. The following is the procedure used to stain DNA with 4.6-diamidino-2-phenylindole (DAPI).

1. After step 4 above, suspend the cell pellet in 1 ml of a staining solution which contains DAPI (Molecular Probes) at a final concentration 1 μg/ml in PBS. Keep on ice for 20 min.

2. Analyse cells by flow cytometry.
   - use excitation with UV light (e.g. 351 nm line from an argon ion laser, or mercury lamp with a UG1 filter)
   - measure the blue fluorescence of DAPI in a band from 460 to 500 nm

Cell analysis by LSC

To be analysed by LSC the cells should be fixed, rinsed with phosphate–citrate buffer, and stained in suspension, as described above, then placed on microscopic slides under a coverslip for measurement. Alternatively, the cells may be cytocentrifuged, smeared, or attached electrostatically to a microscope slide (see Section 2). To attach cells to a microscope slide by cytocentrifugation follow the steps as below:

1. Add 300 μl of cell suspension in tissue culture medium (with serum) containing approximately 20 000 cells into a cytospin (e.g. Shandon Scientific) chamber. Cytocentrifuge at 1000 r.p.m. ($\sim$110 g) for 6 min.

2. Without allowing the cytospun cells to dry completely, fix them by immersing the slides in a Coplin jar containing 70% ethanol. The cells can be stored in ethanol for several days. After fixation, rinse the slides in phosphate–citrate buffer, stain the cells with PI as described above for cell suspensions, and measure their fluorescence by LSC.

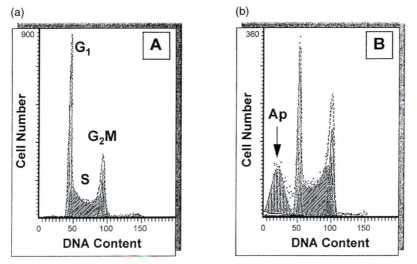

Figure 5. Detection of apoptotic cells with fractional DNA content based on cellular DNA content analysis. To induce apoptosis the cells were treated with DNA topoisomerase II inhibitor, fostriecin, as described (37). The cells were then fixed in 70% ethanol, stained with PI and their fluorescence was measured by flow cytometry (FACScan), as described in *Protocol 4*. The population of apoptotic cells is represented by the sub-G1 peak (Ap).

Apoptotic cells have a decreased PI (or DAPI) fluorescence and diminished forward light scatter compared with the cells in the main peak (G1) (*Figure 5*). It should be emphasized that the degree of extraction of low MW DNA from apoptotic cells, and consequently the content of DNA remaining in the cell for flow cytometric analysis, may vary dramatically depending on the degree of DNA degradation (duration of apoptosis), the number of cell washings, the pH, and the molarity of the washing and staining buffers. Therefore, in *Protocol 4*, step 3, add less or no extraction buffer (e.g. 0–0.2 ml) if DNA degradation in apoptotic cells is extensive (late apoptosis), and more (up to 1.0 ml) if DNA is not markedly degraded (early apoptosis) and there are problems with separating apoptotic cells from G1 cells due to their overlap on DNA content frequency histograms.

7. Identification of apoptotic cells based on the presence of DNA strand breaks

DNA fragmentation during apoptosis generates a large number of DNA strand breaks. The 3′-OH termini in the strand breaks can be detected by attaching a fluorochrome to them. This is generally done by using, directly or indirectly (e.g. via biotin or digoxygenin), fluorochrome-labelled deoxynucleotides in a reaction preferentially catalysed by exogenous TdT, as originally described in refs 14 and 16 and discussed in Chapter 2. Of all the markers of DNA strand breaks, BrdUTP appears to be the most advantageous with respect to sensitivity, low cost, and simplicity of the reaction (17). This deoxynucleotide, once incorporated into DNA strand breaks, is detected by an FITC-conjugated anti-BrdU antibody. It should be stressed that detection of DNA strand breaks requires cell pre-fixation with a cross-linking agent such as formaldehyde, which, unlike ethanol, prevents the extraction of small sections of the fragmented DNA. Thus, despite cell permeabilization (with ethanol) and the subsequent cell washings during the procedure, the DNA content of early apoptotic cells (and with it the number of DNA strand breaks) is not markedly diminished compared with unfixed cells.

Protocol 5. Detection of DNA strand breaks

Reagents

- prepare fixatives—1st fixative: 1% methanol-free formaldehyde (available from Polysciences Inc., Warrington, PA) in PBS, pH 7.4; 2nd fixative: 70% ethanol
- the TdT reaction buffer (5 ×) contains: potassium (or sodium) cacodylate, 1 M; Tris–HCl, 125 mM, pH 6.6; bovine serum albumin (BSA), 1.25 mg/ml
- cobalt chloride ($CoCl_2$), 10 mM
- TdT in storage buffer, 25 units in 1 µl
  The buffer, TdT, and $CoCl_2$ are available from Boehringer Mannheim, Indianapolis, IN.

- BrdUTP stock solution: BrdUTP (Sigma) 2 mM (100 nmoles in 50 µl) in 50 mM Tris–HCl, pH 7.5
- FITC-conjugated anti-BrdU mAb solution (per 100 µl of PBS): 0.3 µg of anti-BrdU FITC-conjugated mAb (available from Becton Dickinson); 0.3% Triton X-100; 1% BSA
- rinsing buffer. Dissolve in PBS: Triton X-100, 0.1% (v/v); BSA, 5 mg/ml
- PI staining buffer. Dissolve in PBS: PI, 5 µg/ml; DNase-free RNase A, 200 µg/ml

Method

1. Fix cells in suspension in 1% formaldehyde for 15 min on ice.

2. Centrifuge (300 *g*, 5 min), resuspend the cell pellet in 5 ml PBS, centrifuge (300 *g*, 5 min), resuspend the cells (approximately 10^6 cells) in 0.5 ml of PBS.

3. Add the above 0.5 ml aliquot of cell suspension into 5 ml of ice-cold 70% ethanol. The cells can be stored in ethanol, at −20°C for several weeks.

4. Centrifuge (200 *g*, 3 min), remove ethanol, resuspend cells in 5 ml PBS, centrifuge (300 *g*, 5 min).

5. Resuspend the pellet (not more than 10^6 cells) in 50 µl of a solution which contains:

   - 10 µl of the reaction buffer
   - 2.0 µl of BrdUTP stock solution
   - 0.5 µl (12.5 units) of TdT in storage buffer
   - 5 µl of $CoCl_2$ solution
   - 33.5 µl distilled H_2O

6. Incubate cells in this solution for 40 min at 37°C (alternatively, incubation can be carried at 22–24°C overnight).

7. Add 1.5 ml of the rinsing buffer, centrifuge (300 *g*, 5 min).

8. Resuspend cells in 100 µl of FITC-conjugated anti-BrdU mAb solution.

9. Incubate at room temperature for 1 h or at 4°C overnight. Add 2 ml of rinsing buffer, centrifuge (300 *g*, 5 min).

10. Resuspend the cell pellet in 1 ml of PI staining solution containing RNase.

11. Incubate for 30 min at room temperature in the dark.

12. Analyse cells by flow cytometry.
    - illuminate with blue light (488 nm laser line or BG12 excitation filter)
    - measure green fluorescence of FITC–anti-BrdU mAb at 530±20 nm
    - measure red fluorescence of PI at >600 nm

Commercial kits

Phoenix Flow Systems and PharMingen Inc., (San Diego, CA, USA) provide kits to identify apoptotic cells based either on a single-step procedure utilizing TdT and FITC-conjugated dUTP (APO-DIRECT™) or TdT and BrdUTP, as described above (APO-BRDU™). A description of the method, which is nearly identical to the above, is included with the kit. Another kit (ApopTag™), based on two-step DNA strand break labelling with digoxygenin–16-dUTP by TdT, is provided by ONCOR Inc., (Gaithersburg, MD, USA).

Analysis of DNA strand breaks by laser-scanning cytometry (LSC)

1. Add 300 µl of cell suspension in tissue culture medium (with serum) containing approximately 20 000 cells into a cytospin chamber. Cytocentrifuge at 1000 r.p.m. (~110 *g*) for 6 min.

2. Without allowing the cytospins to dry completely, pre-fix cells in 1% formaldehyde in PBS for 15 min on ice.

3. Transfer the slides to 70% ethanol and fix for at least 1 h; the cells can be stored in ethanol for several days.

4–9. Follow steps 4–9 above, as described for flow cytometry. Small volumes (50–100 µl) of the respective buffers, rinses, or staining solutions are carefully layered on the cytospin area of the slides held horizontally. At appropriate times these solutions are removed with a Pasteur pipette (or vacuum suction pipette). Small pieces (2.5 cm × 2.5 cm) of thin polyethylene foil may be layered on slides atop the drops to prevent drying. The incubations should be carried out in a moist atmosphere to prevent drying at any step of the reaction.

Protocol 5. *Continued*

10. Rinse the slide in PBS and mount the cells under a coverslip in a drop of the PI staining solution containing RNase A. If the preparations are to be stored for a longer period of time (hours, days, at 4°C), the coverslips are mounted in a drop of a mixture of glycerol and PI staining solution (9:1).

11. Measure cell fluorescence on LSC.

   • excite fluorescence with 488 nm laser line

   • measure green fluorescence of FITC–anti-BrdU mAb at 530±20 nm

   • measure red fluorescence of PI at >600 nm

Identification of apoptotic cells is based on their intense labelling with FITC–anti-BrdU mAb, which frequently requires use of an exponential scale (logarithmic photomultipliers) for data acquisition and display (*Figure 6*). Simultaneous measurement of DNA content makes it possible to identify the cell cycle position of both cells in apoptotic and non-apoptotic populations.

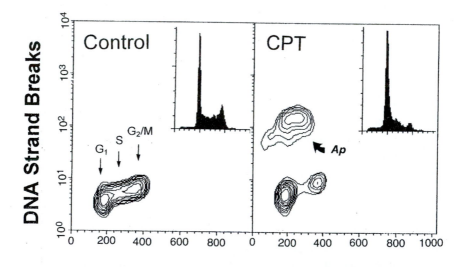

DNA Content

Figure 6. DNA strand break labelling in apoptotic cells. To induce apoptosis HL60 cells were treated with CPT for 3 h as described (16, 17). DNA strand breaks were labelled with BrdUTP and cellular DNA was counterstained with PI, as described in *Protocol 5*. Cellular fluorescence was measured by flow cytometry (FACScan). Note that, preferentially, S phase cells are undergoing apoptosis.

8. Identification of apoptotic cells based on the increased DNA sensitivity to denaturation

The sensitivity of DNA *in situ* to denaturation is higher in the condensed chromatin of apoptotic cells compared with DNA in the interphase nuclei of live cells (37). There is also evidence that a fraction of DNA in apoptotic cells is denatured, reacting with antibody against single-stranded DNA (38). Measurement of DNA denaturability is based on the metachromatic property of the fluorochrome acridine orange (AO) which differentially stains double-stranded (ds) vs. denatured, single-stranded (ss) nucleic acids. In this method, cells are briefly pre-fixed in formaldehyde followed by ethanol. RNA is then removed by pre-incubation with RNase A and DNA is denatured *in situ* by brief cell exposure to 0.1 N HCl. The cells are then stained with AO at pH 2.6; the low pH of the staining reaction prevents DNA renaturation (37). Apoptotic cells, having a larger fraction of DNA in the denatured form, have more intense red and reduced green fluorescence, compared with non-apoptotic (interphase) cells, which stain intensively green but have low red fluorescence (*Figure 7*).

Compared with non-apoptotic interphase cells, apoptotic cells show increased red and decreased green fluorescence (*Figure 7*). Because the increased DNA sensitivity to denaturation is associated with chromatin condensation and seems to be unrelated to the presence of DNA strand breaks (37) this method may be preferable in situations of atypical apoptosis, when internucleosomal DNA degradation does not occur (19–21).

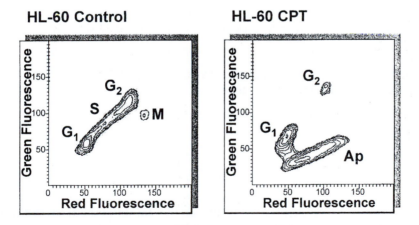

Figure 7. Detection of apoptotic cells based on the increased DNA denaturability. Apoptosis of HL60 cells was induced by CPT, as described (37).The cells were stained with AO after denaturation of DNA with 0.1 M HCl, as described in *Protocol 6*. Their fluorescence was measured by flow cytometry (FACScan). Apoptotic cells (Ap) are characterized by the increased red fluorescence and lowered green fluorescence.

Protocol 6. Identification of apoptotic cells by DNA denaturation

Reagents

Stock solutions, buffers:

- stock solution of AO: dissolve 1 mg of AO (Molecular Probes, Inc.) in 1 ml of distilled water (AO stock solution is stable for several months when stored at 4°C and in the dark)
- RNase stock solution: dissolve 1 mg of RNase A in 1 ml of distilled water (use DNase-free RNase). Less pure preparations require 2–3 min heating at 100°C to inactivate DNase.

- fixatives: (a) 1% methanol-free formaldehyde dissolved in PBS, pH 7.4; (b) 70% ethanol
- HCl, 0.1 M
- AO staining solution: (a) mix 90 ml of 0.1 M citric acid with 10 ml of 0.2 M Na_2HPO_4; the final pH is 2.6; (b) add 0.6 ml of the AO stock solution (1 mg/ml) to 100 ml of this buffer; the final AO concentration is 6 μg/ml. This solution is stable for several weeks when stored in the dark, at 4°C.

Method

1. Rinse the cells with PBS and suspend 10^6–10^7 cells in 1 ml of PBS.

2. Fix the cells by transferring 1 ml of the above cell suspension into a tube containing 9 ml of 1% formaldehyde in PBS, on ice. After 15 min, centrifuge (300 *g*, 5 min).

3. Suspend the cell pellet in 10 ml of PBS, centrifuge (300 *g*, 5 min).

4. Suspend the cell pellet in 1 ml of PBS, transfer this suspension with a Pasteur pipette into a tube containing 9 ml of 70% ethanol, on ice. Samples can be stored in ethanol at 4°C for a minimum of 4 h to up to several weeks.

5. Centrifuge (200 *g*, 3 min) and suspend the cell pellet (10^6–10^7 cells) in 1 ml of PBS.

6. Add 0.2 ml of RNase A stock solution. Incubate at 37°C for 30 min.

7. Centrifuge and resuspend cells in 1 ml of PBS.

8. Withdraw a 0.2 ml aliquot of cell suspension in PBS and transfer it to a small (e.g. 5 ml volume) tube.

9. Add 0.5 ml of 0.1 M HCl, at room temperature.

10. After 30 sec, add 2.0 ml of AO solution, also at room temperature (**Important**: *do not use cold solutions in steps 9 and 10*).

11. Transfer this suspension to the flow cytometer and measure cell fluorescence.

    - AO is excited with blue light; use the 457 (preferred) or 488 nm laser lines, or BG12 excitation filter

    - measure the green fluorescence at 530±20 nm

    - measure red fluorescence with a long pass filter at >640 nm

---

9. General comments

The selection of a particular method depends on the cell system, the nature of the inducer of cell death, the mode of cell death, the information that is being sought (e.g. specificity of apoptosis with respect to the cell cycle phase or DNA ploidy), and the technical restrictions (e.g. the need for sample transportation, type of flow cytometer available, etc.). Positive identification of apoptotic cells is not always easy. The most specific methods appear to be those based on detection of DNA strand breaks. The number of DNA strand breaks in apoptotic cells is so large that the intensity of their labelling seems to be an adequate marker for their identification (14). The situation, however, is complicated in the instances of apoptosis where internucleosomal DNA degradation does not occur (19–21). The number of DNA strand breaks in such atypical apoptotic cells may be inadequate for their identification by this method. Conversely, false-positive identification of apoptotic cells can take place when internucleosomal DNA degradation accompanies cell necrosis.

Apoptosis can be recognized with greater certainty when more than a single viability assay is used. The assays may include simultaneous assessment of plasma membrane integrity, exposure of phosphatidylserine to fluoresceinated annexin V, mitochondrial transmembrane potential, DNA denaturability, and/or DNA fragmentation . For example, preservation of plasma membrane integrity combined with chromatin condensation and extensive DNA breakage is a more specific marker of apoptosis than each of these features alone. However, regardless of the assay used to identify apoptosis, the mode of cell death should be positively identified by inspection of cells by light or electron microscopy. Morphological changes during apoptosis have a very specific pattern (24) and should be the deciding factor in situations where ambiguity arises regarding the mechanism of cell death.

There are characteristic features of LSC distinguishing it from flow cytometry which offer some advantages and even unique technical possibilities in its use in the analysis of cell death. The major advantage of LSC is that cells may be relocated after they are measured, and examined visually or by image analysis. This can confirm the mode of cell death, which, as discussed above, is not possible based on measurement of a single cell attribute. This feature of LSC is useful, for example, in recognizing non-apoptotic phagocytic cells that have ingested apoptotic bodies of neighbouring cells (2). Such phagocytes could be erroneously identified as apoptotic cells by flow cytometry. The possibility of relocating once measured cells on the slide also enables one to perform additional examinations after bleaching and restaining with other fluorochrome(s) or absorbing dyes (39). Since the slides can be stored indefinitely, the same cells can be re-examined in the future using new probes. Since the data, including the cell position on the slide, are stored in list-mode fashion, the measurements performed at different time points on the same cells can be correlated with each other.

Although imaging by LSC is at a lower resolution compared with classical image analysis, LSC offers the possibility of extracting morphometric features from the examined cell that are not available by flow cytometry. The characteristic condensation of chromatin resulting in hyperchromicity of DNA in apoptotic cells, revealed by high maximal pixel values, provides a marker identifying apoptotic cells (30; *Figure 2*). Cell and nuclear area are measured by the number of pixels recording cytoplasmic (e.g. protein) and nuclear (DNA) fluorescence, and the nucleus to cytoplasm ratio can be calculated and can be a valuable parameter of morphological changes that occur during apoptosis.

The duration of apoptosis is usually short (1–6 h) and variable depending on cell type, inducer of apoptosis, *in vitro* vs. *in vivo* conditions, etc. Both prolongation and increased frequency of apoptosis are reflected by the increased apoptotic index . One has to be careful, therefore, in interpreting data that show an increased apoptotic index as representing increased frequency of apoptosis. Likewise, the 'time window' through which a particular flow cytometric method sees this kinetic event, and thus can identify the apoptotic cell, also varies. Hence, the apoptotic index in the same cell population detected by different methods may not be identical. Quantitative estimation of cell death in cell populations requires a simultaneous measure of the absolute number of live cells, in addition to quantitation of apoptosis or necrosis (40).

Acknowledgements

This work was supported by NCI grant CA RO1 28704, the Chemotherapy Foundation and 'This Close' Foundation for Cancer Research. Elzbieta Bedner, the recipient of an Alfred Jurzykowski Foundation fellowship, was on leave from the Department of Pathology, Pomeranian School of Medicine, Szczecin, Poland.

References

1. Darzynkiewicz, Z., Bruno, S., Del Bino, G., Gorczyca, W., Hotz, M.A., Lassota, P., and Traganos, F. (1992). *Cytometry*, **13**, 795–808.
2. Darzynkiewicz, Z., Juan, G., Li, X., Gorczyca, W., Murakami, T., and Traganos, F. (1997). *Cytometry*, **27**,1–20.
3. Frey, T. (1995). *Cytometry*, **21**, 265–274.
4. Ormerod, M.G. (1998). *Leukemia*, **12**,1013–1025.
5. van Engeland, M., Nieland, L.J.W., Ramaekers, F.C.S., Schutte, B., and Reutelingsperger, P.M. (1998). *Cytometry*, **31**, 1–9.
6. Telford, W.G., King, L.E., and Fraker, P.J. (1994). *J. Immunol. Meth.*, **172**, 1–16.
7. Cossarizza, A., Ceccarelli, D., and Masini, A. (1996). *Exp. Cell Res.*, **222**, 84–94.
8. Zamzani, N., Brenner, C., Marzo, I., Susin, S.A., and Kroemer, G. (1998). *Oncogene*, **16**, 2265–2282.

9. Ormerod, M.G., Collins, M.K.L., Rodriguez-Tarduchy, G., and Robertson, D. (1992). *J. Immunol. Meth.*, **153**, 57–66, .

10. Koopman, G., Reutelingsperger, C.P.M., Kuijten, G.A.M., Keehnen, R.M.J., Pals, S.T., and van Oers, M.H.J. (1994). *Blood*, **84**, 1415–1420.

11. Petit, P.X., LeCoeur, H., Zorn, E., Dauguet, C., Mignotte, B., and Gougeon, M.L.J. (1995). *Cell Biol.*, **130**, 157–165.

12. Nicoletti, I., Migliorati, G., Pagliacci, M.C., Grignani, F., and Riccardi, C. (1991). *J. Immunol. Meth.*, **139**, 271–280.

13. Umansky, S.R., Korol', B.R., and Nelipovich, P.A. (1981). *Biochim. Biophys. Acta*, **655**, 281–290.

14. Gorczyca, W., Bruno, S., Darzynkiewicz, R., Gong, J., and Darzynkiewicz, Z. (1992). *Int. J.Oncol.*, **1**, 639–648.

15. Gold, R., Schmied, M., Rothe, G., Zischler, H., Breitschopf, H., Wekerle, H., and Lassman, H. (1993). *J. Histochem. Cytochem.*, **41**, 1023–1030.

16. Gorczyca, W., Gong, J., and Darzynkiewicz, Z. (1993). *Cancer Res.*, **52**, 1945–1951.

17. Li, X. and Darzynkiewicz, Z. (1995). *Cell Prolif.*, **28**, 571–579.

18. Li, X., Melamed, M.R., and Darzynkiewicz, Z. (1996). *Exp. Cell Res.*, **222**, 28–37.

19. Catchpoole, D.R. and Stewart, B.W. (1993). *Cancer Res.*, **53**, 4287–4296.

20. Ormerod, M.G., O'Neill, C.F., Robertson, D., and Harrap, K.R. (1994). *Exp. Cell Res.*, **211**, 231–237.

21. Zamai, L., Falcieri, E., Marhefka, G., and Vitale, M. (1996). *Cytometry*, **23**, 303–311.

22. Hara, S., Halicka, H.D., Bruno, S., Gong, J., Traganos, F., and Darzynkiewicz, Z. (1996). *Exp Cell Res.*, **232**, 372–384.

23. Arends, M.J., Morris, R.G., and Wyllie, A.H. (1990). *Am. J. Pathol.*, **136**, 593–608.

24. Kerr, J.F.R., Wyllie, A.H., and Curie, A.R. (1972). *Br. J. Cancer*, **26**, 239–257.

25. Kamentsky, L.A., Burger, D.E., Gershman, R.J., Kamentsky, L.D., and Luther, E. (1997). *Acta Cytol.*, **41**, 123–143.

26. Kamentsky, L.A., and Kamentsky, L.D. (1991). *Cytometry*, **12**, 381–387.

27. Bedner, E., Burfeind, P., Gorczyca, W., Melamed, M.R., and Darzynkiewicz, Z. (1997). *Cytometry*, **29**, 191–196.

28. Bedner, E., Melamed, M.R., and Darzynkiewicz, Z. (1998). *Cytometry*, **33**, 1–9.

29. Bedner, E., Burfeind, P., Hsieh, T-C., Wu, J.M., Augero-Rosenfeld, M., Melamed, M.R., Horowitz, H.W., Wormser, G.P., and Darzynkiewicz, Z. (1998). *Cytometry*, **33**, 47–55.

30. Furuya, T., Kamada, T., Murakami, T., Kurose, A., and Sasaki, K. (1997). *Cytometry*, **29**, 173–177.

31. Luther, E. and Kamentsky, L.A. (1996). *Cytometry*, **23**, 272–278.

32. Schmid, I., Uttenbogaart, C.H., and Giorgi, J.V. (1994). *Cytometry*, **15**, 12–20.

33. Clatch, R.J., Foreman, J.R., and Walloch, J.L. (1998). *Cytometry*, **34**, 3–16.

34. Salzman, G.C., Singham, S.B, Johnston, R.G., and Bohren, C.F. (1990). In *Flow cytometry and sorting* (ed. M.R. Melamed, T. Lindmo, and M.L. Mendelsohn), pp. 81–107. Wiley-Liss, New York.

35. Darzynkiewicz, Z., Traganos, F., Staiano-Coico, L., Kapuscinski, J., and Melamed, M.R. (1982). *Cancer Res.*, **42**, 799–806.

36. Gong, J., Traganos, F., and Darzynkiewicz, Z. (1994). *Anal. Biochem.*, **218**, 314–319.

37. Hotz, M.A.,Gong J., Traganos, F., and Darzynkiewicz, Z. (1994). *Cytometry*, **15**, 237–244.
38. Frankfurt, O.S., Byrnes, J.J., Seckinger, D., and Sugarbaker, E.V. (1993). *Oncol. Res.*, **5**, 37–42.
39 Li, X., and Darzynkiewicz, Z. (1999) *Exp. Cell Res.*, **249**; 404–412.
40. Darzynkiewicz, Z., Bedner, E., Traganos, F., and Murakami, T. (1998). *Human Cell.*, **11**, 3–12.

5

Cell-mediated cytotoxicity and cell death receptors

JAVIER NAVAL and ALBERTO ANEL

1. Overview of pathways for cell-mediated cytotoxicity

T cell-mediated cytotoxicity in short time assays (3–4 h) is achieved through two main mechanisms. A first mechanism, Ca^{2+} dependent, requires the action of the pore-forming protein perforin and a subfamily of serine proteases named granzymes or fragmentins (1). The second mechanism, which is Ca^{2+} independent for its execution, requires the expression of Fas (Apo-1/CD95) at the target cell surface (2), and of the Fas ligand (FasL) at the effector cell surface (3). The fact that the residual cytotoxic activity detected in T cells from perforin knock-out mice could be attributed solely to Fas-based cyto-toxicity underscored the importance of these two lytic mechanisms (4)

Perforin and granzymes are only expressed in specialized killer cells like cytotoxic T lymphocytes (CTL) or natural-killer (NK) cells, where they are stored in secretory granules (5, 6). After the recognition by CTL or NK cells of a virus-infected or a tumour cell, these granules are secreted and the perforin-mediated entry of granzymes into the cytoplasm causes the rapid lysis of target cells (4, 7). Although also implicated in CTL- and NK-induced lysis, the main physiological role of Fas-based cytotoxicity may be the control of peripheral tolerance (8, 9). The implication of Fas-based cytotoxicity in activation-induced cell death (AICD) of T cell hybridomas (10) and of normal T cell blasts (11) has been clearly demonstrated. It has been suggested that tumour necrosis factor-α (TNF) mediates AICD of $CD8^+$ T cells in the absence of Fas expression, while the Fas/FasL system would be the main cause for AICD of $CD4^+$ T cells (12). These results suggest that T cells able to secrete TNF could use this cytokine for the lysis of target cells expressing TNF receptors. It should be noted, however, that TNF-induced toxicity is usually detected only after long incubation times (48–72 h) (12).

1.1 Fas-based cytotoxicity

1.1.1 Induction of FasL expression in CTL

The expression of FasL in CTL is not constitutive and a strong activation signal through the T cell antigen receptor (TCR) is needed (13). This can also be achieved using a combination of the phorbol ester PMA, which activates efficiently protein kinase C (PKC), and the calcium ionophore ionomycin (2). The induction of FasL expression in CTL can be prevented by protein tyrosine kinase inhibitors, and by cyclosporin A, presumably through the inhibition of calcineurin phosphatase (13). Hence, although the execution of Fas-based cytotoxicity occurs in the absence of extracellular Ca^{2+}, the TCR-induced FasL expression is dependent on Ca^{2+} (14). It was later demonstrated that TCR-induced, but not PMA/ionomycin-induced, FasL expression in CTL was dependent on phosphatidylinositol 3-kinase activity (15), and that the transcriptional regulation of FasL expression was dependent on the activation of the transcription factor NF-AT (16), according to the observed inhibition by cyclosporin A.

1.1.2 Mechanism for Fas-induced apoptosis

Fas (Apo-1/CD95) is a type I transmembrane glycoprotein belonging to the TNF receptor superfamily. Fas is expressed in some thymocyte subsets, activated and tumour T and B cells, and in several non-immune tissues. Fas induces apoptosis when cross-linked with agonist mAbs or by its natural ligand, FasL (17). Fas and the TNF receptor share a cytoplasmic domain needed to induce apoptosis that has been termed the 'death domain' (18).

A great deal of research effort has demonstrated the implication of intra-cellular cysteine–proteases with Asp specificity of the interleukin-1β-converting enzyme (ICE) family, later named as caspases, in apoptosis (see Chapter 10), and in particular in Fas-induced cell death (reviewed in refs 19 and 20)). Using the peptide inhibitor, Ac-DEVD-CHO, which inhibits, preferentially, caspases of the CPP32 (caspase-3) subfamily (21), it has been shown that these caspases play a key role in cytotoxicity exerted by anti-Fas mAbs (22), and by FasL-expressing T cell effectors (23). The finding that activation of CPP32-like caspases plays a pivotal role in many types of apoptotic process has led to the proposal that these caspases constitute the apoptotic executioner (24).

The connection between Fas ligation and the activation of the apoptotic executioner has recently been unveiled. Aggregation of Fas through agonist antibodies or by its physiological ligand induces the recruitment to the cross-linked receptors and, through their respective death domains, of a molecule termed FADD (Fas-associated death domain) (25). In addition to the death domain, which allows its interaction with the aggregated receptors, FADD also contains a 'death effector domain' that mediates the recruitment of another molecule composed of two death effector domains and a caspase domain, named FLICE, MACH, Mch5, or caspase-8 (26–28). The recruitment of this molecule induces activation of its protease activity, resulting in the cleavage and activation of CPP32-like proteases, the apoptotic executioner. An alternative pathway implicated in the activation of CPP32-like caspases is the release of pro-apoptotic proteins (cytochrome *c* and the apoptosis-inducing factor, AIF) from mitochondria in the first stages of apoptosis (29, 30). This process takes place through the assembly of a complex that also includes Apaf-1, the mammalian homologue of Ced-4, and caspase-9, which in turn, activates caspase-3 (31) (see also Chapter 1, Fig. 6). Both processes, 'direct' activation of caspase-3 through caspase-8 recruitment to the aggregated receptor and 'indirect' activation of caspase-3 through mitochondrial activation of caspase-9, are not mutually exclusive. In fact, it has recently been demonstrated that the relative importance of one over the other depends on the nature of the cells undergoing apoptosis (32).

1.2 Perforin/granzyme-based cytotoxicity

1.2.1 Perforin and granzymes

The insertion of perforin into the plasma membrane of target cells facilitates the entry of granzymes inside the cell cytoplasm (5, 33). Granzymes, in turn, activate programmed cell death (34). At least 10 different serine proteases are expressed in CTL and/or NK cells, granzyme A and granzyme B being the major components of lytic granules and the main contributors to cytotoxicity (6, 34, 35). Although granzyme B is a serine protease, it shares with caspases an unusual substrate specificity characterized by the requirement of an Asp residue at the P1 position. Granzyme A, on the contrary, has a tryptase-like specificity (34), suggesting functional versatility of the system. This versatility has been evidenced in granzyme A-knock-out mice, where cell-mediated cytotoxicity is normal (36), and in granzyme B-knock-out mice, where although cell-mediated cytotoxicity is blocked at short incubation times (4 h), it is completely recovered at longer times (16 h) (37).

1.2.2 Mechanism for perforin/granzymes-induced apoptosis

A possible connection between granzyme-induced cell death and caspase activation has been demonstrated, since it was observed that purified granzyme B cleaves and activates caspase-3 (38) or its close homologue caspase-7 (39). It was later demonstrated that in short-term assays (3–4 h), the peptide inhibitors Ac-DEVD-CHO (40) and Z-VAD-fmk (41) prevented nuclear apoptosis induced by CTL on Fas-negative targets, as measured by the [125I]iododeoxyuridine release assay and nuclear staining. These studies clearly implicated caspases in perforin/granzyme-based cytotoxicity exerted by CTL. However, caspase inhibitors did not prevent cell lysis under the same or similar experimental conditions, as measured by the ^{51}Cr release assay (41, 42) or membrane manifestations of apoptosis, like phosphatidylserine exposure (41). In addition, the inhibition of CTL-induced nuclear apoptosis

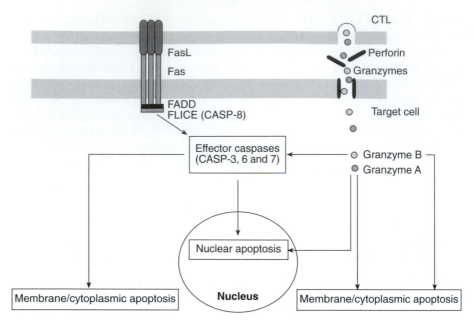

Figure 1. Schematic representation of the known biochemical mechanisms for apoptosis induction by CTL on target cells (see Section 1 for details).

by Ac-DEVD-CHO observed in 3–4 h assays was no longer observed in 6–16 h assays. The combination of Ac-DEVD-CHO and an inhibitor of granzyme A, inhibited nuclear apoptosis at any time-point tested (40). Altogether, these data indicate that: (i) granzyme B induces the nuclear manifestations of apoptosis through a mechanism mediated by caspases; (ii) granzyme B induces cytoplasmic apoptotic manifestations leading to cell lysis through a caspase-independent mechanism; and (iii) all the manifestations of granzyme A-induced apoptosis, nuclear and cytoplasmic, seem to proceed through a caspase-independent mechanism. A summary of all these considerations is shown in *Figure 1*.

2. Standard methods for the evaluation of cell-mediated cytotoxicity

2.1 Chromium release assay

Perhaps the most simple, cheap, and effective technique for *in vitro* cytotoxicity tests is the one based on target cell labelling with $Na_2^{51}CrO_4$. This is probably still the most widely used method and is commonly referred to as 'the chromium (^{51}Cr) release assay'. Radioactive chromate passes through the cell membrane by an unknown mechanism and is bound mainly to cytoplasmic proteins. Since ^{51}Cr is a γ-emitter, no special preparation of cell samples is needed for radioactive determination. This technique was first used to evaluate the effect of radiation on thymocytes (43) and complement-mediated cytotoxicity (44), and was subsequently applied to the study of cell-mediated cytotoxicity (45). This method measures lysis of target cells, so both apoptotic and non-apoptotic deaths are detected.

Protocol 1. Chromium-51 release assay

Safety recommendations

Chromium-51 is a medium-energy γ-radiation emitter and it must be manipulated according to the corresponding national regulations for use of radioactive materials. The researcher must wear appropriate protective clothing (laboratory coat, surgical gloves) and dosimeter. Stock, concentrated solutions of $Na_2{}^{51}CrO_4$ must be conveniently shielded (half-value layer: 3 mm lead, approximately). On the other hand, the manipulation of ^{51}Cr does not require any other special precaution, since in the form of chromate it is not absorbed by any organ in the body.

Equipment and reagents

- tissue culture facilities
- effector and target cells
- complete culture medium (e.g. RPMI 1640 supplemented with heat-inactivated 5% FCS, 5 mM Hepes, pH 7.4, and antibiotics)
- V-shaped 96-well microtitre plates
- sterile $Na_2{}^{51}CrO_4$ in aqueous solution (from Amersham, NEN, or ICN) (1 mCi/ml)
- γ-radiation counter (solid scintillation)

A. *Labelling of target cells*

The protocols given are adequate for labelling of tumour target cells and for lymphocytes from mixed lymphocyte cultures (MLC). Some examples of commonly used murine target cells, including their MHC restriction, are the following:

- L1210: lymphocytic leukaemia-expressing H-2^d
- EL-4: thymoma-expressing H-2K^b
- RMA: lymphoma-expressing H-2K^b
- P815: monocytic leukaemia-expressing H-2^d
- L929: fibrosarcoma-expressing H-2K^k (adherent cell)

Two labelling alternatives are possible.

(a) *Rapid (2 h) labelling.*

1. Resuspend the appropriate amount of washed, cultured targets in non-diluted FCS at 1×10^6 cells in 100 μl.

2. Add 10 μCi of $Na_2{}^{51}CrO_4$ per 1×10^6 cells and incubate at 37°C for 1 h.

Protocol 1. *Continued*

3. Wash then by dilution and centrifugation in complete medium, re-suspend at 1×10^6 cells/ml in complete medium, and incubate for another hour at 37 °C.

4. Eliminate the second supernatant, wash again in complete medium, and resuspend the cells for the experiment.

(b) *Overnight labelling*

1. Resuspend around 2×10^6 target cells in 5 ml of complete medium, place the suspension in a 25 cm² culture flask, and add 200–300 μCi of $Na_2{}^{51}CrO_4$.

2. Incubate overnight at 37 °C in a CO_2 incubator, and then wash twice with RPMI 1640.

3. Resuspend the cells in complete medium for the experiment.

B *Cytotoxicity test*

1. The manipulation during the cytotoxicity tests can be done under non-sterile conditions. Resuspend the labelled target cells at 10^5 cells/ml and add 100 μl per well of 96-well plates, so that 10^4 labelled target cells are used per experimental point. Make points at least in triplicate.

2. Wash and resuspend the effector cells in complete medium. Several effector to target (E:T) ratios should be used. In the case of CTL clones, E:T ratios that should give a good lysis are between 1:1 and 10:1. In the case of primary mixed lymphocyte cultures (MLC), the E:T ratios recommended are between 20:1 and 100:1. Resuspend the effector cells at the maximum cell density to be used in a given experimental point. Then, plate 100 μl on the wells already containing the labelled target cells, previously having made the corresponding dilutions of the effectors with complete medium to obtain the E:T ratios chosen.

3. Include at least triplicates of (i) labelled target cells alone resuspended in 200 μl of complete medium; and (ii) labelled target cells alone in 100 μl of complete medium where 100 μl of 1% Triton X-100 or 2 N HCl are added to estimate, respectively, the spontaneous release and total labelling of the targets.

4. Centrifuge the plate at 400 *g* at room temperature for 1 min to favour cell–cell contact. Incubate at 37 °C in a cell incubator for the time chosen. Usually, 4 h cytotoxicity tests are performed, but longer times can also be used, provided that spontaneous release of the label from the targets does not increase above 20% of the total labelling. If spontaneous release in the experimental conditions used is greater than 30%, the assay is not reliable.

5. At the end of the incubations, centrifuge the plates at 400 *g* during 5 min and harvest 100 μl aliquots of each supernatant by using a multichannel pipette. Transfer the aliquots directly to small RT-15 plastic tubes (Bibby Sterilin), seal them with molten paraffin wax, and determine the radioactivity associated with the supernatants in the γ-counter.

6. Calculate the specific ^{51}Cr release at the different E:T ratios as follows:

% of lysis = % of specific ^{51}Cr release =

$$\frac{\text{(cpm sample–cpm spontaneous release)}}{\text{(cpm maximum release–cpm spontaneous release)}} \times 100.$$

2.2 [^{125}I]Iododeoxyuridine (^{125}IUdR) release assay

^{125}IUdR competes with thymidine for incorporation during DNA synthesis because the iodine atom is sterically similar to the methyl group of thymidine. So ^{125}IUdR is incorporated into DNA and, when the target cell is killed by effector cells, ^{125}IUdR is released in the form of DNA fragments that are not re-utilized. Since this event is a nuclear marker of apoptosis, ^{125}IUdR release is the method of choice to analyse nuclear apoptosis during cell-mediated cytotoxicity. This method was initially used by comparing it with ^{51}Cr release to distinguish CTL-induced death from lysis induced by antibody plus complement, which is not associated with DNA fragmentation (46, 47).

Protocol 2. [^{125}I]Iododeoxyuridine (^{125}IUdR) release assay

Precautions

^{125}I is a low-energy γ- emitter, and is more easily shielded (half-value layer: 0.02 mm lead) than ^{51}Cr. The same precautions as indicated for ^{51}Cr manipulation should be used here (*Protocol 1*). The incorporation of ^{125}I into the thyroid by inhalation is its main biological risk. However, since ^{125}IUdR is not volatile, special precautions are not needed for its manipulation.

Equipment and reagents

- tissue culture facilities
- effector and target cells
- sterile 1.5 ml Eppendorf cones
- culture medium (e.g. RPMI 1640 supplemented with heat-inactivated 5% FCS, 5 mM Hepes, pH 7.4, and antibiotics)
- sterile ^{125}IUdR (from Amersham, NEN, or ICN) (1 mCi/ml)
- hypotonic lysis buffer: 0.5% Triton X-100 in 20 mM Tris, 1 mM EDTA, pH 7.5
- γ-counter (solid scintillation counter)

A. *Labelling of target cells*

1. It is convenient for target tumour cells to be in an exponential phase of growth when labelling with ^{125}IUdR.

Protocol 2. *Continued*

2. Resuspend the washed target cells in complete medium at 5×10^5 cells/ml.

3. Add 10–15 μCi of ^{125}IUdR per ml and incubate in a CO_2 incubator at 37°C for 2–3 h. Then, wash the cells twice with RPMI 1640 and resuspend in complete medium for the experiment.

B. *Cytotoxicity test*

1. Resuspend the labelled target cells at 1×10^5 cells/ml and add 100 μl to a 1.5 ml Eppendorf tube, so that 10^4 labelled target cells are used per experimental point. Each experimental point should be made in triplicate at least.

2. Wash and resuspend the effector cells in complete medium. The same considerations about E:T ratios indicated for the ^{51}Cr release assay (*Protocol 1*) apply here. Resuspend the effector cells at the maximum cell density to be used in a given experimental point. Then, add 100 μl into the tubes already containing the labelled target cells, having previously made the corresponding dilutions of the effectors with complete medium to obtain the E:T ratios chosen.

3. Include duplicates, at least, of labelled target cells alone, resuspended in 200 μl of complete medium to estimate the spontaneous release of the label from the targets.

4. Centrifuge the Eppendorf tubes at 400 *g* at room temperature for 2 min to favour cell–cell contact. Incubate at 37°C in a cell incubator for the time chosen. Usually, 4 h cytotoxicity tests are performed, but longer times can also be used, provided that spontaneous release of the label from the targets is not greater than 20% of the total labelling.

5. Centrifuge the tubes at 400 *g* for 5 min, harvest carefully the whole supernatant, and save it for counting.

6. Lyse the cells by adding 500 μl of ice-cold lysis buffer, followed by brief vortexing, and incubation on ice for 20 min. Centrifuge the lysates at 12 000 *g* for 10 min at 4°C in a minifuge, collect the supernatants containing fragmented DNA and cut off (with scissors or blade) the bottom of the tube that contains non-fragmented DNA.

7. Determine the radioactivity associated with each sample in a γ-counter, including the incubation medium, the supernatant of the lysate, and the DNA pellet.

8. Calculate the specific ^{125}IUdR release at the different E:T ratios as follows:

% DNA fragmentation = % of specific ^{125}IUdR release =

$$\frac{(\text{cpm sample} - \text{cpm spontaneous})}{(\text{cpm total} - \text{cpm spontaneous})} \times 100$$

cpm sample = (cpm in the incubation medium + cpm in the cell lysate) of the sample

cpm spontaneous = (cpm in the incubation medium + cpm in the cell lysate) spontaneous

cpm total = (cpm in the incubation medium + cpm in the cell lysate + cpm in the DNA pellet) of the sample

2.3 The JAM test

This method measures, using target cells labelled with [³H]thymidine and a conventional cell-harvester, the intact DNA that remains associated with living target cells. This allows for the estimation of the DNA amount that has undergone fragmentation during cell-mediated cytotoxicity (48). Hence, the JAM test detects DNA fragmentation induced by effector cells on the labelled targets and the data obtained can be considered equivalent to those obtained using the ^{125}IUdR release assay.

Protocol 3. [³H]Thymidine-labelled DNA retention by living cells

Precautions

³H radioactivity is a low-energy β-radiation emitter. Conventional plastic and glass containers and surgical gloves protect quite well from this type of radiation. On the other hand, due to its low energy, tritium β-particles must be detected in a liquid scintillation counter, making the sample preparation slightly more laborious. This method is fast, sensitive, and easy to perform if an automated cell-harvester is available, but it generates more radioactive waste.

Equipment and reagents

- tissue culture facilities
- effector and target cells
- sterile round-bottom, 96-well plates
- culture medium (e.g. RPMI 1640 supplemented with heat-inactivated 5% FCS, 5 mM Hepes, pH 7.4, and antibiotics)
- sterile [³H]methylthymidine (from Amersham, NEN, or ICN) (1 mCi/ml)
- automated cell-harvester
- scintillation vials and liquid scintillation fluid
- β-counter (liquid scintillation counter)

A *Labelling of target cells*

1. It is very convenient if the target cells to be in an exponential phase of growth when labelling with [³H]thymidine.

2. The day before the assay, resuspend the target cells into fresh medium and add 1 ml to the wells of a 24-well tissue culture plate at 3–5 × 10⁵ cells/ml.

Protocol 3. *Continued*

3. To label, add [³H]thymidine to a final concentration of 5 μCi/ml to the cultures and incubate under the same culture conditions for 4–6 h. After the incubation, wash the cells twice with RPMI 1640 and resuspend in complete medium for the experiment.

B *Cytotoxicity test*

1. Resuspend labelled target cells at 10^5 cells/ml and add 100 μl per well, so that 10^4 labelled target cells are used per experimental point. Each experimental point should be made in triplicate at least.

2. Wash and resuspend the effector cells in complete medium. The same considerations about E:T ratios indicated for the previously described assays also apply here. Resuspend the effector cells so that 100 μl contains the maximum effector number to be used in a given experimental point. Add 100 μl of the corresponding dilution of effector cells to the labelled target cells, in order to obtain the desired E:T ratios.

3. Include at least triplicates of labelled target cells alone, resuspended in 200 μl of complete medium to estimate the spontaneous release of label from the targets. To calculate the total labelling, a triplicate of labelled target cells can be directly placed into scintillation vials.

4. Centrifuge the plates at 400 *g* at room temperature for 2 min to favour cell–cell contact. Incubate at 37°C in a CO_2 incubator for the time chosen (1–4 h).

5. At the end of the incubations, aspirate the cells and medium from the plate wells on to fibre glass filters using an automated cell-harvester.

6. Wash the filters, dry, and count them in a liquid scintillation counter.

7. Calculate the percentage of DNA fragmentation at the different E:T ratios as follows:

% specific DNA fragmentation =

$$\frac{(\text{spontaneous cpm} - \text{sample cpm})}{\text{spontaneous cpm}} \times 100$$

spontaneous cpm = cpm from retained labelled DNA in the absence of killers

sample cpm = cpm from retained labelled DNA in the presence of killers

Although the total incorporation (total labelling of target cells in the absence of effector cells) is not needed for the calculation of specific DNA loss, it does need to be measured in each assay, to be sure that the DNA retained in the absence of killers ('spontaneous cpm') is roughly the same as the total labelling.

2.4 BLT–esterase release assay

One of the consequences of CTL activation through the antigen receptor or by the combination of phorbol esters and calcium ionophores is the exocytosis of cytoplasmic granules containing perforin and granzymes. The presence of these proteins in the supernatant of stimulated CTL can be detected by a colorimetric method termed the BLT–esterase release assay (49). Trypsin-type serine esterases are detected in fact in these assay, granzyme A being predominant in CTL or NK cells.

Protocol 4.   BLT–esterase release assay

Equipment and reagents

- tissue culture facilities
- effector and target cells
- culture medium (e.g. RPMI 1640 supplemented with heat-inactivated 5% FCS, 5 mM Hepes, pH 7.4, and antibiotics)
- flat and round-bottomed 96-well plates
- ELISA plate reader

- benzyloxycarbonyl-L-lysine thiobenzyl ester (BLT) (from Sigma or Calbiochem) and dithiobis-(2-nitrobenzoic) acid (DTNB) (from Sigma). BLT powder should be stored at –20°C inside a desiccator, while DTNB can be stored at room temperature.

Assay

1. Wash and resuspend the effector cells in complete medium. Stimulate the cells with the appropriate stimulus. The stimuli could vary: anti-TCR/CD3 antibodies, phorbol esters plus calcium ionophores, alloantigen-bearing target cells, etc. If target cells are chosen as stimulants, several effector to target (E:T) ratios may be used. In this case, the E:T ratios recommended are 1:1, 1:2, and 1:3, increasing the number of targets, not of effectors, to optimize granule release from the CTL. In this case, resuspend effector cells in complete medium at 2×10^6 cells/ml and add 50 μl of the suspension (1×10^5 cells) into the wells of a round-bottomed 96-well plate. Resuspend the target cells so that 50 μl contains the maximum target cell number to be used in a given experimental point and then make the corresponding dilutions with complete medium. Plate 50 μl of the target cell suspensions on the wells that already contain the CTL, in order to obtain the E:T ratios chosen.

2. Include at least triplicates of (i) effector cells alone resuspended in 100 μl of complete medium; and (ii) effector cells alone in 50 μl of complete medium where 50 μl of 1% Triton X-100 are added to estimate, respectively, the spontaneous release and total CTL content of BLT–esterase.

3. Centrifuge the plate at 400 *g* at room temperature for 2 min to favour cell–cell contact. Incubate at 37°C in a cell incubator during 3–4 h.

Protocol 4. *Continued*

4. At the end of the incubations, centrifuge the plate at 400 *g* for 10 min, harvest carefully 30 µl of the supernatants, and plate them on the wells of a clean, flat-bottomed, 96-well plate. Make a triplicate blank with 30 µl of fresh culture medium.

5. During the last centrifugation, prepare the BLT reagent by dissolving 4 mg of BLT and 4 mg of DTNB in 50 ml of phosphate-buffered saline, pH 7.4, by gently shaking and vortexing. Add 150 µl of the reagent to each well of the plate and incubate at 37°C. The BLT reagent is of limited stability and can only be used for 3 or 4 h after preparation.

6. Read the intensity of the yellow colour formed using an ELISA plate reader at 405–415 nm, depending on the filters available, subtracting automatically the blank values from the sample values. It is recommended that absorbances be read after different incubation times, for example a first reading at 15 min, followed by measurements at 30 min, 45 min, and 1 h.

7. Calculate the specific esterase release at the different E:T ratios as follows :

% of specific BLT–esterase release =

$$\frac{[(A_{405})\ \text{sample}-(A_{405})\ \text{spontaneous}] \times 100}{[(A_{405})\ \text{total}-(A_{405})\ \text{spontaneous}]}$$

2.5 Estimation of target cell nuclear fragmentation using fluorescent dyes

Apoptosis is associated with chromatin condensation and nuclear fragmentation. By using fluorescent dyes that label the DNA, these morphological nuclear changes can be detected by using fluorescence microscopy (see Chapters 2 and 4 for a more detailed discussion). We will describe here briefly a method based on *p*-phenylenediamine (PPDA) labelling (23), but other fluorescent dyes, like Hoechst 33342, have been successfully used (41) [we recommend chapter 2 and the work of Nakajima *et al.* (50) where the use of Hoechst 33342 is described]. PPDA is currently used as solution in pure glycerol as a mounting, anti-fading reagent, but when prepared in oxidized glycerol, it gives a fluorescent adduct which stains nuclei.

Protocol 5. Estimation of target cell nuclear fragmentation by PPDA staining

Contrary to other nuclear fluorochromes, the yellow to orange PPDA fluorescence can be visualized with the same filter arrangement as for FITC, the most commonly used.

Equipment and reagents

- tissue culture facilities
- effector and target cells
- culture medium (e.g. RPMI 1640 supplemented with heat-inactivated 5% FCS, 5 mM Hepes, pH 7.4, and antibiotics)
- 3-(4,5-dimethylthiazol-2-yl)-2,5-diphenyl-tetrazolium bromide (MTT)
- 1% paraformaldehyde in PBS (fixation buffer)
- round cover glasses (13 mm diameter) previously treated for 20 min with a 0.1 mg/ml solution of poly-L-lysine (Sigma) in water

- sterile Eppendorf cones
- PPDA mounting medium: 10 mg PPDA dissolved in 1 ml PBS, pH 7.4, and then mixed with 9 ml of oxidized glycerol. Neutralize if necessary with 1 M NaOH. To prepare oxidized glycerol, treat 10 ml glycerol with 100 μl H_2O_2 (33%, 110 vols), aliquot, and store for 1 day in a crystal-clear tube, exposed to sunlight at room temperature. Once prepared, store the PPDA reagent until use in the dark at −20°C.
- fluorescence microscope (excitation filter: 450–490 nm)

Assay

1. PPDA staining is done after the fixation of cells, and so target cells can be previously incubated with MTT to distinguish them from effector cells. For that, incubate target cells with MTT at 0.2 mg/ml in complete medium for 3–4 h under normal culture conditions. Formazan crystals will appear inside the cells, in the form of microspikes, that can be visualized in the microscope under white light. These spikes do not interfere with the cytotoxic assay.

2. Wash the target cells and resuspend in complete medium at 1×10^6 cells/ml. Add 100 μl to Eppendorf tubes so that 1×10^5 target cells are used per experimental point.

3. Wash and resuspend the effector cells in complete medium. Several E:T ratios can be used but it is recommended not to go over a 5:1 ratio, since the presence of too many effectors in the images could hinder the finding of apoptotic target cells.

4. Include a control with target cells alone resuspended in 200 μl of complete medium to check for spontaneous apoptosis during the assay.

5. Centrifuge the Eppendorf tubes at 400 *g* at room temperature for 5 min to favour cell–cell contact. Incubate at 37°C in a CO_2 incubator for the time required.

6. At the end of incubations, centrifuge the tubes at 400 *g* for 10 min and

Protocol 5. *Continued*

discard supernatants. Wash the cells with RPMI 1640 and fix in 1% paraformaldehyde in PBS, pH 7.4, at room temperature for 15 min.

7. Place the poly-L-lysine pre-treated, round cover glasses into wells of 24-well plates and drop upon the fixed cell suspensions. Centrifuge the plates for 5 min at 200 *g* to fix the cells to the glass cover.

8. Wash (twice) the cover glasses containing the fixed cells with PBS, air-dry, and finally place each cover glass over a drop (20 μl) of PPDA reagent previously added to glass slides.

9. Visualize fluorescent-labelled nuclei using a fluorescence microscope. Non-apoptotic cells will predominate (effectors), but they can be easily distinguished from targets using normal illumination by the absence of formazan crystals. In non-apoptotic cells, nuclei are stained uniformly, giving an orange-red fluorescence of medium intensity. Apoptotic nuclei are characterized by chromatin condensation, which gives a much brighter yellow PPDA fluorescence, accompanied by the characteristic chromatin condensation and nuclear fragmentation. The number of apoptotic cells can be evaluated by counting different fields and photographing.

2.6 Activation-induced cell death (AICD)

The so-called AICD refers to the process by which an overactivated T cell kills itself or its neighbours. This process occurs physiologically during peripheral T cell deletion and is important for the termination of the immune response and the prevention of autoimmune attack. AICD is known to be mediated by the interactions between FasL/Fas, APO2L/DR4, TNF/TNFR, and possibly other uncharacterized molecules (51). AICD may be induced *in vitro* by incubation of T cell blasts, leukaemic or hybridoma cells with mitogenic lectins [phytohaemagglutinin (PHA), concanavalin A], phorbol esters, or anti-TCR/CD3 antibodies. The peculiarity of the process lies in the fact that effector cells are also the target cells and, therefore, the classical methods for evaluating cell-mediated cytotoxicity using radioisotopes (^{51}Cr, ^{125}IUdR) are not applicable. For this type of assay it is recommended that a suitable dye-reduction method is used (e.g. the MTT assay; see ref. 52) to analyse cell death induced by a given stimulus. If bystander killing needs to be evaluated, a cytotoxicity assay using the methods described in *Protocols 1* and *2* should be performed. In this case, stimulated T cells should be used as effectors, while non-stimulated T cells of the same type should be used as targets.

3. Separate studies of Fas- and perforin/granzyme-based cytotoxicity

3.1 Fas-based cytotoxicity in the absence of perforin/granzyme contribution

One of the best ways to study Fas-based cytotoxicity in the absence of the perforin/granzyme contribution is using cytolytic effector cells that are defective in the perforin/granzyme system, while keeping intact their ability to induce FasL expression. This was first done in the laboratory of Dr Pierre Golstein (Marseille, France) by serial subcloning from the cytolytic hybridoma PC60 to generate d10S, d11S, and d12S cells (2). These cells were able to kill Fas-expressing cells once activated with a combination of PMA plus ionomycin whether in the presence or absence of extracellular calcium. They were later used in the laboratory of Dr Shikegazu Nagata (Osaka, Japan) to clone FasL (3).

Another approach has been to isolate effector cells from perforin knock-out mice. In these mice, the whole perforin/granzyme system is defective, while FasL expression is normal (1, 4).

However, normal effector cells, harbouring all the cytotoxicity mechanisms mentioned, can also be used to study Fas-based cytotoxicity in the absence of a perforin/granzyme contribution, or vice versa. This allows a good estimation of the relative contribution of each cytotoxicity mechanism in a given experimental system. We will describe suitable methods using whole functional CTL clones (13, 15).

Protocol 6. Assay of Fas-based cytotoxicity in the absence of perforin/granzyme contribution

Reagents

- 10 ng/ml PMA
- 600 nM ionomycin
- anti-TCR or anti-CD3 antibodies
- sodium carbonate/bicarbonate buffer, pH 8.0
- 1 mM EGTA
- 1.5 mM $MgCl_2$

Method

1. Culture CTL clones in complete culture medium supplemented with IL-2, in 24-well plates, to a density between 3 and 6×10^5 cells/ml. Then, stimulate the cells for 3 h in the same culture conditions to induce FasL expression with one of the following stimuli:

   (i)   a combination of PMA and ionomycin.

   (ii) anti-TCR or anti-CD3 antibodies. In the case of anti-CD3 antibodies, their previous immobilization on wells of a 24-well plate

---

Protocol 6. *Continued*

by overnight incubation of a 20 μg/ml antibody solution at 4°C in sodium carbonate/bicarbonate buffer recommended.

(iii) antigen-bearing target cells. In this case, it is recommended that a 1:0.7 effector to target ratio is used, to minimize the interference between the activating cells and the Fas-bearing target cells in the subsequent cytotoxicity assay.

2. After stimulation, wash out the stimulating agents, and test functional FasL expression in a cytotoxicity test on Fas-expressing target cells using one of the protocols described above. A good system for such an assay is the use of a cell line that does not express Fas and its corresponding Fas transfectant. For instance, the leukaemic cell line L1210 has been transfected with the Fas cDNA, generating an identical cell line which expresses high Fas membrane levels and is very sensitive to Fas-induced apoptosis (2). Then, Fas-based cytotoxicity can be assayed using any pre-stimulated effector cell on L1210Fas cells and using L1210 cells as a negative control. Any other system with such characteristics can also be chosen (e.g. L929/L929Fas).

3. To avoid any hypothetical contribution of perforin/granzymes, cyto-toxicity tests on L1210 and L1210Fas cells should be performed in the absence of extracellular calcium. To do this, add EGTA and MgCl$_2$ to the medium of the cytotoxicity test. The execution of Fas-based cyto-toxicity, once FasL is expressed, is calcium independent, while the execution of perforin/granzyme-based cytotoxicity is entirely calcium dependent.

3.2 Perforin/granzyme-based cytotoxicity in the absence of FasL contribution

In this case, a suitable option would be to test the cytotoxicity of effector cells on antigen-bearing target cells that do not express Fas. Murine L1210 cells, which express H-2K^d, have been used by several authors to estimate perforin/granzyme-based cytotoxicity of anti-H-2K^d effectors (42, 53). However, perforin/granzyme-based cytotoxicity exerted by effector cells of any antigen specificity can also be tested using anti-CD3 antibody-redirected lysis on L1210 cells (40).

Protocol 7.   Anti-CD3 antibody-redirected lysis of Fas-negative target cells

Reagents

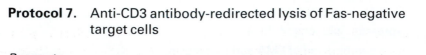

- L1210 cells
- 3 μg/ml anti-CD3 antibody
- 1 mM EGTA
- 1.5 mM MgCl$_2$

Method

1. Effector cells should be treated in the same way as in a normal cytotoxicity assay. Targets of choice for this assay should be Fas-negative and express Fc receptors, conditions which are met by L1210 cells. Label the target cells with ^{51}Cr or ^{125}IUdR, as described in *Protocols 1* and *2*, and perform the assay as indicated, but including anti-CD3 antibody in the medium of the cytotoxicity test.

2. To ensure that all the cytotoxicity observed is due to perforin/granzyme, the inclusion of a negative control with EGTA and MgCl$_2$ in the medium of the cytotoxicity test is recommended.

4. Use of caspase and granzyme inhibitors in cell-mediated cytotoxicity assays

Initial studies on caspase and granzyme inhibitors were done *in vitro*, using purified molecules and cell-free systems. The validation of the results obtained concerning the possible implication of several caspases and/or granzymes in cell-mediated cytotoxicity should come using whole-cell assays. Several peculiarities of the inhibitors and of the assays should be kept in mind when performing these studies, which we will briefly consider now.

Editors note: Additional cautions regarding the use of caspase inhibitors are discussed in Chapter 10, Section 8.2.

4.1 Caspase inhibitors

The most frequently used caspase inhibitors are the following:

- Ac-YVAD-cmk or Ac-YVAD-CHO, specific inhibitors for caspase-1 (ICE) (54, 55)
- Ac-DEVD-fmk or Ac-DEVD-CHO, which preferentially inhibit at low concentrations caspases of the CPP32-like subfamily, especially caspase-3 (CPP32) and caspase-7 (ICE-LAP3) (21). At higher concentrations they may also inhibit caspase-1, caspase-8 (FLICE), and caspase-10 (Mch4) (55).
- Z-VAD-fmk, a general caspase inhibitor (56), although its ability to inhibit caspases of the CPP32-like subfamily and caspase-2 (Ich-1) is limited (55).
- Z-VDVAD-fmk, a specific inhibitor of caspase-2, which needs a 5 amino acid peptide to be efficiently inhibited (55). It also inhibits caspase-3, though less efficiently than DEVD.
- Boc-D-fmk, a general caspase inhibitor (57)

Although granzyme B is also a protease with Asp specificity, peptide fluoromethylketones inhibit caspases but not granzymes. To inhibit serine proteases, the more reactive peptide chloromethylketones are needed (57).
Several biotechnological companies sell these peptides, among them

Bachem (Bubendorf, Switzerland), Neosystem Laboratoire (Strasbourg, France), and Enzyme System Products (Dublin, California).

Protocol 8. Use of caspase inhibitors in whole-cell cytotoxicity assays

Method

1. Stock solutions of peptide caspase inhibitors are normally prepared in DMSO and stored at −20°C. Recommended concentrations of the stocks range between 100 and 300 mM. Peptides are better stored lyophilized at 4°C.

2. In many cases, a pre-incubation of target cells with peptide caspase inhibitors is needed to ensure enough incorporation by the cells. Z-VAD-fmk, Z-VDVAD-fmk, and Boc-D-fmk are effective at concentrations between 20 and 100 μM and 1 h of pre-incubation is recommended. Ac-DEVD-CHO has three lateral carboxyl chains, hindering its passage through the cell membrane. Therefore, in spite of being a very efficient caspase-3 inhibitor *in vitro* (K_i <1 nM), high concentrations (between 300 and 600 μM) and pre-incubation times of 3 h are needed to guarantee that the inhibitory intracellular concentration is reached. Some biotechnological companies have solved this permeability problem by methylation of the carboxylic side chains, e.g. the commercially available Ac-DEVD-fmk. In this case, the efficient concentration to be used and the time of pre-incubation recommended is similar to that of Z-VAD-fmk.

3. Another consideration should be made when using this type of inhibitor, and especially the non-methylated Ac-DEVD-CHO, regarding the effector to target ratio. If there is a great excess of effector cells, then the inhibitor will be taken up by the effectors, greatly reducing the actual peptide concentration in the medium and, consequently, lowering its intracellular concentration inside the targets. For example, no effect of Ac-DEVD-CHO is observed on Fas-based cytotoxicity when using an E:T ratio of 10:1, while it is completely inhibited at an E:T ratio of 5:1 (23). Hence, E:T ratios from 1:1 to 5:1, but not greater, should be used in this type of experiment.

4. When using any inhibitor in a cytotoxicity assay, the possible increase in spontaneous release of the label from target cells induced by the inhibitor itself should be carefully controlled. Each experimental point where an inhibitor is included should be accompanied by a spontaneous release control of target cells alone, incubated with the inhibitor at the same concentration as in the assay. If the spontaneous release induced by the inhibitor alone is greater than 30% of the total

labelling, then this inhibitor cannot be used in such an assay. In our hands, peptide caspase inhibitors do not induce increases in spontaneous release either of ^{51}Cr or ^{125}IUdR, even at high concentrations (up to 1 mM) and long incubation times (up to 16 h).

5. Bearing in mind these considerations, perform any of the cytotoxicity tests described in *Protocols 1* and *2* in the presence of peptide caspase inhibitors.

4.2 Granzyme inhibitors

The chemistry of granzymes has been extensively studied in the laboratory of Dr James Powers (Atlanta, Georgia) in collaboration with several other groups (see ref. 58). For example, the BLT–esterase release method described in Section 2.4 is one of the applications of these studies. As discussed above, peptide inhibitors of serine proteases should be chloromethylketones. However, these compounds, because of their high reactivity, could inhibit other cellular proteases. For this reason, several other chemical inhibitors, mainly isocoumarin derivatives, have been developed.

- For granzyme B, the most specific peptide inhibitor described is Z-AAD-cmk, which does not inhibit granzyme A or related proteases (6). In the case of isocoumarins, 3,4-dichloroisocoumarin (DCI) has been described to inhibit granzyme B much more efficiently than granzyme A (around 50-fold difference in K_i) (59).

- Granzyme A, unlike granzyme B, is a trypsin-like protease. In a cytotoxicity assay, it can be considered that the major targets for these inhibitors are trypsin-like granzymes, granzyme A being the predominant one. The most specific peptide inhibitor for trypsin-like proteases is d-FPR-cmk, which does not inhibit granzyme B (6). On the other hand, 7-(phenyl-ureido)-4-chloro-3-(2-isothioureidoethoxy)-isocoumarin, abbreviated as IGA, is the most potent isocoumarin described for granzyme A inhibition (60, 61).

Z-AAD-cmk and d-FPR-cmk are sold by Enzyme System Products (Dublin, California) and DCI is sold by many chemical companies (e.g. Sigma).

Protocol 9.   Use of granzyme inhibitors in whole-cell cytotoxicity assays

Method

1. Stock solutions of peptide and iosocumarin granzyme inhibitors are usually prepared in DMSO and stored at –20 °C.

2. Pre-incubation of inhibitors with target cells is not needed in this case

---

Protocol 9. *Continued*

since the inhibitors exhibit sufficient permeability through the cell membrane. The effective concentrations for DCI, IGA, and Z-AAD-cmk are between 20 and 50 μM.

3. The considerations of effector to target ratios in *Protocol 8* also apply here. Hence, E:T ratios from 1:1 to 5:1, but not greater, should be used in this type of experiment.

4. The possible increase in spontaneous release of the label from target cells induced by the inhibitor itself should be carefully controlled. If the spontaneous release induced by the inhibitor alone is greater than 30% of the total labelling, then this inhibitor cannot be used in such an assay. In our hands, Z-AAD-cmk, DCI and IGA do not induce an increase in the spontaneous release in ^{125}IUdR release assays at the concentrations mentioned, even when using long incubation times (up to 16 h). However, these inhibitors do induce spontaneous ^{51}Cr release in long-term assays. These inhibitors can only be used at the indicated concentrations in short-term assays (3–5 h), and the particular conditions of incubation with each target cell used need to be carefully optimized.

5. Bearing in mind these considerations, perform any of the cytotoxicity tests described above in the presence of the described granzyme inhibitors.

Acknowledgements

We would like to acknowledge the high-quality scientists from whom we learned the techniques described above, and who transmitted to us their passion for cell-mediated cytotoxicity. In chronological order, thanks to Alan Kleinfeld, Anne O'Rourke, Matt Mescher, Michel Buferne, Claude Boyer, Anne-Marie Schmitt-Verhulst, and Pierre Golstein.

References

1. Kägi, D., Ledermann, B., Bürki, K., Seiler, P., Odermatt, B., Olsen, K. J., Podack, E. R., Zinkernagel, R. M. and Hengartner, H. (1994). *Nature*, **369**, 31.
2. Rouvier, E., Luciani, M. F. and Golstein, P. (1993). *Journal of Experimental Medicine*, **177**, 195.
3. Suda, T., Takahashi, T., Golstein, P. and Nagata, S. (1993). *Cell*, **75**, 1169.
4. Kägi, D., Vignaux, F., Ledermann, B., Bürki, K., Depraetere, V., Nagata, S., Hengartner, H. and Golstein, P. (1994). *Science*, **265**, 528.
5. Henkart, P. A., Millard, P., Reynolds, C. W. and Henkart, M. P. (1984). *Journal of Experimental Medicine*, **160**, 75.

6. Shi, L., Kam, C. M., Powers, J. C., Aebersold, R. and Greenberg, A. H. (1992). *Journal of Experimental Medicine*, **176**, 1521.
7. Van den Broek, M. F., Kägi, D., Ossendorp, F., Toes, R., Vamvakas, S., Lutz, W. K., Melief, C. J. M., Zinkernagel, R. M. and Hengartner, H. (1996). *Journal of Experimental Medicine*, **184**, 1781.
8. Russell, J. H., Rush, B., Weaver, C. and Wang, R. (1993). *Proceedings of the National Academy of Sciences USA*, **90**, 4409.
9. Vignaux, F. and Golstein, P. (1994). *European Journal of Immunology*, **24**, 923.
10. Dhein, J., Walczak, H., Bäumler, C., Debatin, K. M. and Krammer, P. H. (1995). *Nature*, **373**, 438.
11. Alderson, M. R., Tough, T. W., Davis-Smith, T., Braddy, S., Falk, B., Schooley, K. A., Goodwin, R. G., Smith, C. A., Ramsdell, F. and Lynch, D. H. (1995). *Journal of Experimental Medicine*, **181**, 71.
12. Zheng, L., Fisher, G., Miller, R. E., Peschon, J., Lynch, D. H. and Lenardo, M. J. (1995). *Nature*, **377**, 348.
13. Anel, A., Buferne, M., Boyer, C., Schmitt-Verhulst, A. M. and Golstein, P. (1994). *European Journal of Immunology*, **24**, 2469.
14. Vignaux, F., Vivier, E., Malissen, B., Depraetere, V., Nagata, S. and Golstein, P. (1995). *Journal of Experimental Medicine*, **181**, 781.
15. Anel, A., Simon, A. K., Auphan, N., Buferne, M., Boyer, C., Golstein, P. and Schmitt-Verhulst, A. M. (1995). *European Journal of Immunology*, **25**, 3381.
16. Latinis, K. M., Carr, L. L., Peterson, E. J., Norian, L. A., Eliason, S. L. and Koretzky, G. A. (1997). *Journal of Immunology*, **158**, 4602.
17. Nagata, S. and Golstein, P. (1995). *Science*, **267**, 1449.
18. Tartaglia, L. A., Ayres, T. M., Wong, G. H. W. and Goeddel, D. V. (1993). *Cell*, **74**, 845.
19. Nagata, S. (1997) *Cell*, **88**, 355.
20. Nicholson, D. W. and Thornberry, N. A. (1997). *Trends in Biochemical Sciences*, **22**, 299.
21. Nicholson, D. W., Ali, A., Thornberry, N. A., Vaillancourt, J. P., Ding, C. K., Gallant, M., Gareau, Y., Griffin, P. R. *et al.* (1995). *Nature*, **376**, 37.
22. Schlegel, J., Peters, I., Orrenius, S., Miller, D. K., Thornberry, N. A., Yamin, T. T. and Nicholson, D. W. (1996). *Journal of Biological Chemistry*, **271**, 1841.
23. Anel, A., Gamen, S., Alava, M. A., Schmitt-Verhulst, A. M., Pineiro, A. and Naval, J. (1996). *International Immunology*, **8**, 1173.
24. Henkart, P. A. (1996). *Immunity*, **4**, 195.
25. Chinnaiyan, A. M., O'Rourke, K., Tewari, M. and Dixit, V. M. (1995). *Cell*, **81**, 505.
26. Muzio, M., Chinnayian, A. M., Kischkel, F. C., O'Rourke, K., Shevchenko, A., Ni, J., Scaffidi, C., Bretz, J. D., Zhang, M., Gentz, R., Mann, M., Krammer, P. H., Peter, M. E. and Dixit, V. M. (1996). *Cell*, **85**, 817.
27. Boldin, M. P., Goncharov, T. M., Goltsev, Y. V. and Wallach, D. (1996). *Cell*, **85**, 803.
28. Fernandes-Alnemri, T., Armstrong, R. C., Krebs, J., Srinivasula, S. M., Wang, L., Bullrich, F., Fritz, L. C., Trapani, J. A., Tomaselli, K. J., Litwack, G. and Alnemri, E. S. (1996). *Proceedings of the National Academy of Sciences USA*, **93**, 7464.
29. Susin, S. A., Zamzami, N., Castedo, M., Hirsch, T., Marchetti, P., Macho, A., Daugas, E., Geuskens, M. and Kroemer, G. (1996). *Journal of Experimental Medicine*, **184**, 1331.

30. Liu, X., Kim, C. N., Yang, J. and Wang, X. (1996). *Cell*, **86**, 147.
31. Li, P., Nijhawan, D., Budihardjo, I., Srinivasula, S. M., Ahmad, M., Alnemri, E. S. and Wang, X. (1997). *Cell*, **91**, 479.
32. Scaffidi, C., Fulda, S., Srinivasan, A., Friesen, C., Li, F., Tomaselli, K. J., Debatin, K. M., Krammer, P. H. and Peter, M. E. (1998). *EMBO Journal*, **17**, 1675.
33. Liu, C. C., Walsh, C. M. and Young, J. D. E. (1995). *Immunology Today*, **16**, 194.
34. Smyth, M. J. and Trapani, J. A. (1995). *Immunology Today*, **16**, 202.
35. Pasternack, M. S. and Eisen, H. N. (1985). *Nature*, **314**, 743.
36. Ebnet, K., Hausmann, M., Lehmann-Grube, F., Müllbacher, A., Kopf, M., Lamers, M. and Simon, M. M. (1995). *EMBO Journal*, **14**, 4230.
37. Heusel, J. W., Wesselschmidt, R. L., Shresta, S., Russell, J. H. and Ley, T. J. (1994). *Cell*, **76**, 977.
38. Darmon, A.J., Nicholson, D.W. and Bleackley, R.C. (1995). *Nature*, **377**, 446.
39. Gu, Y., Sarnecki, C., Fleming, M. A., Lippke, J. A., Bleackley, R. C. and Su, M. S. S. (1996). *Journal of Biological Chemistry*, **271**, 10816.
40. Anel, A., Gamen, S., Alava, M. A., Schmitt-Verhulst, A. M., Pineiro, A. and Naval, J. (1997). *Journal of Immunology*, **158**, 1999.
41. Sarin, A., Williams, M. S., Alexander-Miller, M. A., Berzofsky, J. A., Zacharchuk, C. M. and Henkart, P. A. (1997). *Immunity*, **6**, 209.
42. Darmon, A. J., Ley, T. J., Nicholson, D. W. and Bleackley, C. R. (1996). *Journal of Biological Chemistry*, **271**, 21709.
43. Scaife, J. F. and Vittorio, P. V. (1964). *Canadian Journal of Biochemistry*, **42**, 503.
44. Sanderson, A. R. (1964). *British Journal Experimental Pathology*, **45**, 398.
45. Holm, G. and Perlham, P. (1967). *Immunology*, **12**, 525.
46. Oldman, R. K. and Herberman, R. B. (1973). *Journal of Immunology*, **111**, 1862.
47. Russell, J. H., Masakowski, V. R. and Dobos, C. B. (1980). *Journal of Immunology*, **124**, 1100.
48. Matzinger, P. (1991). *Journal of Immunological Methods*, **145**, 185.
49. Takayama, H., Trenn, G., Humphrey, W., Bluestone, J. A., Henkart, P. A. and Sitkovsky, M. V. (1987). *Journal of Immunology*, **138**, 566.
50. Nakajima, H., Lichtenfels, R., Martin, R. and Henkart, P. A. (1995). In *Techniques in apoptosis: a user's guide* (ed. T.G. Cotter and S.J. Martin), pp. 175–190. Portland Press, London.
51. Ashkenazi, A. and Dixit, V. M. (1998). *Science*, **281**, 1305.
52. Skehan, P. (1999). In *Cell growth, differentiation, and senescence: a practical approach* (ed. G.P. Studzinski). Oxford University Press, Oxford.
53. Simon, M. M., Hausmann, M., Tran, T., Ebnet, K., Tschopp, J., Thahla, R. and Müllbacher, A. (1997). *Journal of Experimental Medicine*, **186**, 1781.
54. Thornberry, N. A., Bull, H. G., Calaycay, J. R., Chapman, K. T., Howard, A. D., Kostura, M. J. *et al.* (1992). *Nature*, **356**, 768.
55. Talanian, R. V., Quinlan, C., Trautz, S., Hackett, M. C., Mankovich, J. A., Banach, D., Ghayur, T., Brady, K. D. and Wong, W. W. (1997). *Journal of Biological Chemistry*, **272**, 9677.
56. Pronk, G.J., Ramer, K., Amiri, P. and Williams, L.T. (1996). *Science*, **271**, 808.
57. Sarin, A., Wu, M. L. and Henkart, P. A. (1996). *Journal of Experimental Medicine*, **184**, 2445.
58. Powers, J. C. and Kam, C. M. (1994). In *Methods in enzymology* (ed. A. J. Barrett), Vol. 244, p. 442. Acadaemic Press, New York.

59. Odake, S., Kam, C. M., Narasimhan, L., Poe, M., Blake, J. T., Krahenbuhl, O., Tschopp, J. and Powers, J. C. (1991). *Biochemistry*, **30**, 2217.
60. Kam, C. M., Kerrigan, J. E., Plaskon, R. R., Duffy, E. J., Lollar, P., Suddath, F. L. and Powers, J. C. (1994). *Journal of Medicinal Chemistry*, **37**,1298.
61. Beresford, P. J., Kam, C. M., Powers, J. C. and Lieberman, J. (1997). *Proceedings of the National Academy of Sciences USA*, **94**, 9285.

6

Sphingolipids as messengers of cell death

GARY M. JENKINS and YUSUF A. HANNUN

1. Introduction

This chapter discusses the most commonly used methods for the study of sphingolipids in their role as messengers of cell death. A basic review of sphingolipids and the role they play in apoptosis is given. The rest of the chapter outlines the methods used for the study of sphingolipids and the enzymes relevant to the sphingomyelin (SM) cycle. Each method is described with full experimental details, and emphasis is given to one or two protocols used for the study of each class of sphingolipids and relevant enzymes.

2. Sphingolipids and their role in apoptosis

2.1 Overview of sphingolipids

Mammalian cells have four major classes of sphingolipids, which are the sphingoid backbones, ceramides, sphingomyelins, and glycosphingolipids (*Figure 1*). Sphingoid backbones are the basic building blocks of sphingolipids formed by the condensation of serine and palmitoyl CoA. Mammalian cells have two predominant forms of sphingoid backbones, which are sphingosine and dihydrosphingosine, each composed of an 18 carbon backbone. The addition of an N-linked fatty acid to the sphingoid backbone creates ceramides. The fatty acid in mammalian cells is usually a 16–24 carbon structure and can be attached to either sphingosine or dihydrosphingosine. Recent research indicates that dihydroceramide is formed first by acylation of dihydrosphingosine and then desaturated to form ceramide (1). Next, an addition of a phosphatidylcholine-derived phosphocholine creates the sphingomyelins. Mammalian cells can have a wide variety of sphingomyelins depending on the ceramides present. The final and predominant class of sphingolipids are the glycosphingolipids formed by various additions of sugar groups to the ceramides.

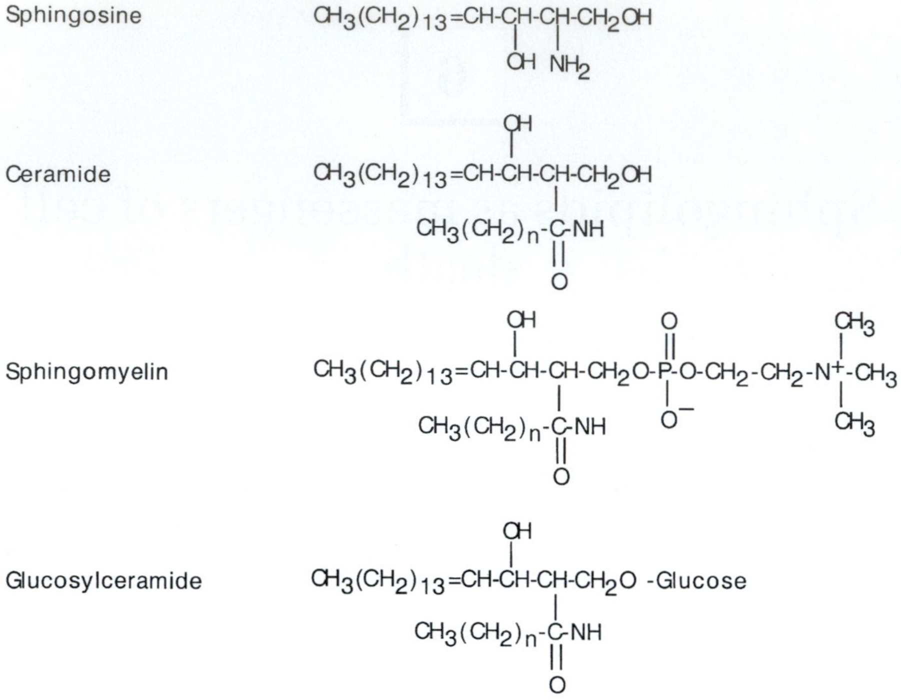

Sphingosine

$$CH_3(CH_2)_{13}=CH-CH-CH-CH_2OH$$
$$\phantom{CH_3(CH_2)_{13}=CH-}OH\ NH_2$$

Ceramide

$$\phantom{CH_3(CH_2)_{13}=CH-CH-}OH$$
$$CH_3(CH_2)_{13}=CH-CH-CH-CH_2OH$$
$$\phantom{CH_3(CH_2)_{13}=CH-}CH_3(CH_2)_n-C-NH$$
$$\phantom{CH_3(CH_2)_{13}=CH-CH-CH-CH_2}O$$

Sphingomyelin

$$\phantom{CH_3(CH_2)_{13}=CH-}OHOCH_3$$
$$CH_3(CH_2)_{13}=CH-CH-CH-CH_2O-P-O-CH_2-CH_2-N^+-CH_3$$
$$\phantom{CH_3(CH_2)_{13}=CH-}CH_3(CH_2)_n-C-NHO^-CH_3$$
$$\phantom{CH_3(CH_2)_{13}=CH-CH-CH-CH_2}O$$

Glucosylceramide

$$\phantom{CH_3(CH_2)_{13}=CH-CH-}OH$$
$$CH_3(CH_2)_{13}=CH-CH-CH-CH_2O\ -Glucose$$
$$\phantom{CH_3(CH_2)_{13}=CH-}CH_3(CH_2)_n-C-NH$$
$$\phantom{CH_3(CH_2)_{13}=CH-CH-CH-CH_2}O$$

Figure 1. Examples of each class of sphingolipids in mammalian cells. The classes represented are, in order from top to bottom, sphingoid backbones, ceramides, sphingomyelins, and glycosphingolipids.

2.2 Role of sphingolipids in apoptosis

The role of sphingolipids in the process of apoptosis is centred on the sphingo-myelin (SM) cycle (*Figure 2*). The inducers of the sphingomyelin cycle include many agents that induce apoptosis and/or growth arrest in cells, and examples are: cytokines such as TNF-α, interleukin-1, and γ-interferon; Fas ligand; 1,25-dihydroxyvitamin D_3; and environmental stresses such as ultraviolet radiation, serum withdrawal, and chemotherapeutic agents (2).

The initial finding pointing to sphingolipids in apoptosis was the observa-tion that ceramide was often cytotoxic to U937 cells, resulting in DNA frag-mentation (3), while closely related compounds such as dihydroceramide and diacylglycerol did not induce DNA fragmentation. Furthermore, recent studies have shown that ceramide acts as a lipid mediator in cells and has, for targets, a ceramide-activated protein kinase and a ceramide-activated protein phos-phatase. Downstream of these effects are the caspases, Raf-1 and the ERKs (2, 4) (*Figure 2*). The generation of ceramide is thought to be from the hydrolysis of SM by the neutral sphingomyelinase (n-SMase) (5) and/or acid

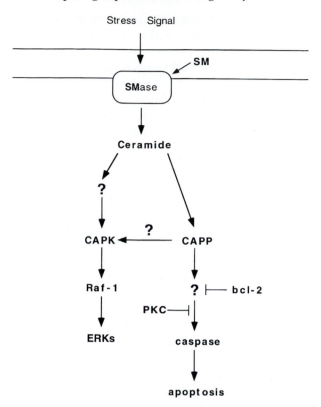

Figure 2. Overview of the role of the sphingomyelin cycle in apoptosis. A stress signal activates sphingomyelinase (SMase) which cleaves sphingomyelin (SM), forming ceramide. Ceramide can induce a ceramide-activated protein phosphatase (CAPP) and/or a ceramide-activated protein kinase (CAPK). CAPP effects can be blocked by Bcl-2 or PKC and leads to activation of caspases, and apoptosis. CAPK works through a different pathway, activating Raf-1 and then the extracellular signal regulated kinases (ERKs).

sphingomyelinase (a-SMase). Furthermore, the overexpression of Bcl-2, a known protector against apoptosis, has been able to protect cells from ceramide-induced apoptosis. Therefore, the study of sphingolipids in the process of apoptosis has become an important field.

2.3 Strategies and considerations in evaluating sphingolipids

Efficient study of sphingolipids in cellular responses entails several basic considerations. The selection of appropriate time points is one of these considerations. Time points should be selected to be relevant to when the biology to be studied is observed. Often this means time points in the range of hours in many cases of sphingolipids in apoptosis; however, shorter time points can be informative for quick changes in sphingolipids. Another basic consideration is the use of time-matched controls from which to compare values from treated samples.

The analysis of various sphingolipid levels requires an initial step for extraction of lipids followed by mild base hydrolysis, which helps by removing most of the glycerolipids whose ester bonds are sensitive to this form of hydrolysis. Quantitation of specific sphingolipids can then be performed using specific and specialized assays. Finally, the levels of sphingolipids are adjusted to either total lipid levels or to cell number.

The starting point for analysis of sphingolipids in apoptosis is the measurement of ceramide levels. When analysing ceramide levels using the diacylglycerol kinase assay, care should be exercised in using the enzyme in a quantitative range and not a catalytic range. The reason for starting with ceramide levels is twofold. First, ceramide has been correlated with many cellular events such as differentiation, stress responses, and apoptosis (2). Secondly, changes of ceramide levels are often found in response to stress. An increase in ceramides can result from many pathways, including hydrolysis of sphingomyelin or glycoceramides, inhibition of sphingomyelin synthase or ceramidase, and/or enhancement of *de novo* synthesis of ceramides. Therefore, once an increase in ceramide is found, further analysis of sphingolipids becomes relevant. The measurement of sphingomyelin can show whether the increase in ceramide levels is from its hydrolysis or another source. Sphingomyelin hydrolysis has been implicated in some ceramide increases, as represented by the SM cycle (2). Sphingoid backbone levels have not been implicated to the same extent as sphingomyelin and ceramide levels. Enhancement of sphingolipid turnover studies can be gained by analysis of the SMases and SM synthase. Changes in the activity of these enzymes reflect either an induction in enzyme amounts and/or some poorly defined form of post-translational modification. Neither the acid/neutral SMase assay or SM synthase assay will detect reversible allosteric regulation of the enzymes. Together, the above analyses allow for an initial in depth look at the possible role of sphingolipids in a given cellular response.

3. Extraction of sphingolipids and normalization by lipid phosphate

3.1 Extraction of sphingosines, ceramides, and sphingomyelins

Extraction of sphingosines, ceramides, and sphingomyelins is most easily accomplished via two commonly used methods. The most prevalent is the method of Bligh and Dyer (6). Also of use is the Folch extraction (7). The two methods rely upon the hydrophobic/organic nature of the sphingolipids to separate them from other cellular components (*Figure 3*). The resulting

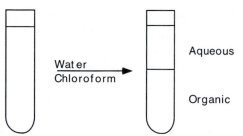

Figure 3. Illustration of the phase break induced in the Bligh and Dyer extraction by the addition of water and chloroform. To obtain clear phases allow to sit or spin in a table-top centrifuge at 2000 g. The organic phase may be collected by first aspirating the aqueous phase, then collecting it, or by directly taking the organic phase.

extract can be used for analysis of sphingolipids using various protocols and these measurements can be normalized to lipid phosphate.

Protocol 1.   Extraction of sphingolipids by the method of Bligh and Dyer (6)

Reagents
- PBS
- chloroform/methanol (1:2 pre-mixed)
- methanol
- dry nitrogen

Method

1. Grow $2–5 \times 10^6$ cells per sample to be analysed. Treat as desired. Adherent cells need to be suspended.

2. Pellet cells by table-top centrifugation of 500 g for 5 min.

3. Aspirate off the media or wash PBS from the pellet.

4. Resuspend the cell pellet in 3 ml of chloroform/methanol (1:2 pre-mixed) and vortex until an even suspension is gained.

5. Add 0.8 ml of water and vortex. Transfer the resuspended cells to a glass screw-cap tube and set on the bench for 30 min or overnight at 4°C for best extraction. If there is a phase break at this point it can be corrected by the addition of 0.5 ml of methanol. A premature phase break can hinder proper extraction.

6. Pellet cellular debris via table-top centrifugation at 2000 g for 5 min. Transfer liquid material to a fresh tube and discard cellular debris.

7. Add 1 ml of chloroform and 1 ml of water and vortex. These additions will induce a break of the liquid material into an organic (lower) and aqueous (upper) phases (*Figure 3*). Allow 15–30 min for the phases to separate and then centrifuge for 5 min at 2000 g to obtain clean phase separation.

Protocol 1. *Continued*

8. One may now collect the lower organic phase directly or aspirate the aqueous phase and then collect the organic phase.
9. The extracted lipids should be in the 2 ml of chloroform and should be dried down via a speed vacuum apparatus or under a stream of dry nitrogen.
10. The dried down lipids should now be resuspended in chloroform with one aliquot designated for phosphate measurement (usually one-third) and another aliquot for the experimental measurement desired (usually one-third; the last third is a back-up).

Protocol 2. Extraction of sphingolipids by the method of Folch

Reagents
- chloroform
- methanol

- dry nitrogen

Method

1. Grow 2–5 × 10^6 cells per sample to be analysed. Treat as desired. Adherent cells need to be suspended.
2. Pellet cells by table-top centrifugation of 500 *g* for 5 min.
3. Aspirate off the media or wash liquid from the pellet.
4. Add 1 ml of methanol to the pellet, vortex, and transfer to a glass screw-cap tube.
5. Add 2 ml of chloroform to the above and vortex. Make sure that the cell pellet is completely resuspended. For best extraction, let samples sit on the bench for 30 min or at 4°C overnight.
6. Pellet cellular debris via table-top centrifugation at 2000 *g* for 5 min. Transfer liquid material to a fresh tube and discard cellular debris.
7. Add 0.8 ml of water and vortex. This addition should induce a phase break in the samples to an organic (lower) and aqueous (upper) phases.
8. One may now collect the lower organic phase directly or aspirate the aqueous phase and then collect the organic phase.
9. The extracted lipids should be in the 2 ml of chloroform and should be dried down via a speed vacuum apparatus or under a stream of dry nitrogen.
10. The dried down lipids should now be resuspended in chloroform with one aliquot designated for phosphate measurement (usually one-third) and another aliquot for the experimental measurement desired (usually one-third; the last third is a back-up).

Once the extraction and aliquoting of lipids is accomplished the actual analysis of each sphingolipid class can begin. Storage of the lipids should be as either a powder or in chloroform at –20°C. As noted in step 10 of either protocol, you should end up with three tubes, each containing one-third of your extract. Both methods are useful for the extraction of the three classes of sphingolipids, but are not efficient for sphingoid phosphates.

3.2 Alkaline hydrolysis of sphingolipids

Analysis of sphingoid backbones and sphingomyelins requires an additional step of refinement. This step is a mild alkaline hydrolysis, which is useful for hydrolysing most glycerolipids, leaving the resistant sphingolipids. Mild alkaline hydrolysis is not used for ceramide measurement when there is additional interest in diacylglycerol, which is sensitive to the procedure. Thus, one can quickly and easily purify the organic lipid extract to one containing mainly sphingolipids.

Protocol 3.   Mild alkaline hydrolysis of lipids

Reagents

- chloroform
- 2 N KOH

- methanol

Method

1. Resuspend dried down lipids from either the Bligh and Dyer (*Protocol 1*) or Folch (*Protocol 2*) extractions in 2 ml of chloroform.

2. Add 0.2 ml of KOH in methanol to the tubes and vortex vigorously.

3. Incubate in a shaking water bath at 37°C for 1 h.

4. Immediately add 0.2 ml of 2 N HCl in methanol to neutralize the KOH.

5. Now add 0.6 ml of methanol and 0.4 ml of water and vortex to break the solution into an organic (lower) and aqueous (upper) phases. The lower phase contains the alkaline-resistant sphingolipids, while the upper has the other hydrolysed lipids.

6. Aspirate off the upper, aqueous phase. Wash the organic phase with 1 ml of pH-neutral water (make neutral by use of 50 μl of 1 N NH$_4$OH per 15 ml of distilled water). Vortex after adding the water and let the samples set for 15 min. Spin at 2000 *g* in a table-top centrifuge to clarify the phase break then aspirate off the aqueous phase. Repeat wash as above.

7. Dry down the organic phase under nitrogen or by a speed vacuum.

3.3 Lipid phosphate measurement for normalization of experimental samples

The comparison of control versus treated samples is of utmost importance in the analysis of *in vivo* sphingolipids. Therefore, a method to ensure consistency between samples is necessary. The most commonly used is that of lipid phosphate measurement taken directly from the aforementioned sphingolipid extraction methods. Furthermore, lipid phosphate can be used to more accurately quantitate sphingolipid samples. This method normalizes for any variance in cell number and extraction efficiency, since lipid phosphate should not vary much between cells. The method relies upon a standard curve analysis and uses a colorimetric assay of phosphate that has been ashed.

Protocol 4.   Measurement of lipid phosphate

Reagents

- 1 mM NaH$_2$PO$_4$
- ashing buffer: H$_2$O:10 N H$_2$SO$_4$:70% HClO$_4$ (40:9:1)
- 0.9% w/v ammonium molybdate
- 10% w/v L-ascorbic acid

Method

1. Prepare a standard curve of phosphates in duplicate. Put 0, 3, 5, 7, 10, 12, 15, 20, and 30 μl of NaH$_2$PO$_4$ (1 nmol. per μl) in labelled tubes.

2. Dry down by speed vacuum or dry nitrogen the aliquots from either extraction above. Use the same style of glass tube for both the standards and the samples to ensure consistent ashing.

3. Add 0.6 ml of ashing buffer and vortex.

4. Place samples in a heating block at 160°C overnight. Approximately 100 μl should be left in the tube after incubation.

5. Add 0.9 ml of water and mix after allowing the samples to cool.

6. Now add 0.5 ml of ammonium molybdate and add 0.2 ml of L-ascorbic acid. Mix tubes very vigorously.

7. Incubate at 45°C for 30 min.

8. Read samples on a spectrophotometer at 820 nm in 0.8 ml cuvettes. First construct the standard curve then use it to get the experimental values.

The number obtained from the phosphate measurement now becomes the denominator for the samples. This means that results are expressed as picomoles of sphingolipids/nanomoles of lipid phosphate. This will be accomplished by comparing the value of the third obtained above to the value of the measurement obtained in the experimental third of sphingolipids. However, phosphate levels should be used to normalize sphingolipids for comparison of cells from a given cell line and not for comparison between cell lines. Cell lines can have very different lipid phosphate levels per cell. Therefore, analysis should be limited to time-matched controls with a change in only one variable.

4. Analysis of sphingoid backbones

4.1 HPLC on sphingoid backbones

The measurement of sphingoid backbones is most prevalently done by derivatization and separation via high performance liquid chromatography (HPLC) (8). This method of analysis requires a mild alkaline hydrolysis of the lipids. The sphingoids are then derivatized on their free amine group (only available on the sphingoid backbones) with *ortho*-phthaldialdehyde (*o*-PA), a fluorescent compound. The samples are then injected and separated on HPLC, yielding a peak area for each sphingoid. The sensitivity of this system is around 5 pmol. This method allows for the separation of sphingoid backbones by their saturation, chain length, and, to a certain extent, stereochemistry (*Figure 4a*).

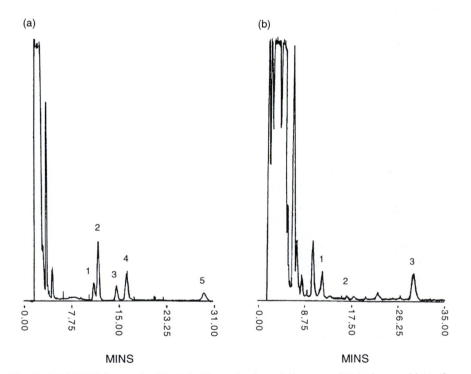

Figure 4. (a) HPLC chart of sphingolipid standards as follows: peak 1, L-threo sphingosine (11.6 min); peak 2, D-erythro sphingosine (12.3 min); peak 3, L-threo dihydrosphingosine (15.2 min); peak 4, D-erythro dihydrosphingosine (16.9 min); peak 5, C20 sphingosine (29.5 min). (b) HPLC chart of sphingoid backbones extracted from MOLT4 cells. Peak 1 is sphingosine (12.3 min), peak 2 is D-erythro dihydrosphingosine (16.5 min), and peak 3 is the internal standard C20 sphingosine (29.2 min). The front peak on each chart is unreacted *o*-PA. Times given are the retention times, which can change with each run due to variation in columns, different back pressures, and slight differences in the running buffer.

113

Stereoisomers can be separated on a reverse phase column, but diastereomers require a chiral column. A further way of comparison available for this analysis is the use of an internal standard. An internal standard should ideally be a non-endogenous sphingoid backbone. Addition of the internal standard allows for comparison of samples within each experiment. 200 pmol of internal standard need to be added to every sample by step 5 of either the Folch (*Protocol 2*) or Bligh and Dyer (*Protocol 1*) extraction given above. In addition to the internal standard, the running of samples of specific sphingoid backbones is wise owing to possible shifts in retention times. Retention times shift because of the different reverse phase C18 columns used, back pressure changes, minor variances in the solvent systems, and type of HPLC system. Knowing the exact retention time of known standards allows for the selection of the proper peaks from mammalian samples (*Figure 4b*) from the background peaks.

Protocol 5. HPLC analysis of sphingoid backbones

Reagents

- D-erythro sphingosine and D,L-erythro dihydrosphingosine (Sigma)
- 5% o-PA
- 3% K-borate buffer (pH 10.5 by addition of KOH pellets to boric acid solution)

- β-mercaptoethanol
- 100% ethanol
- running buffer: 90% methanol, 10% 5 mM potassium phosphate, pH 7.0

Method

1. Prepare 100–200 pmol of known standards of D-erythro sphingosine and D,L-erythro dihydrosphingosine for derivatization. These standards are useful for defining retention times within an HPLC run. If necessary a standard curve of values can be constructed with known standards of the D -erythro sphingosine.

2. On the day of running the samples on HPLC, one should make up the o-PA reagent fresh. This requires 9.9 ml of 3 % K-borate buffer, 5 μl of β-mercaptoethanol and 0.1 ml of 5% o-PA in 100% ethanol (for example, weigh out over 5 mg of o-PA and resuspend in 20 times the volume of weighed material in μl, i.e. 8.6 mg would be resuspended in 172 μl of ethanol).

3. In order to derivatize dried down lipids from the alkaline hydrolysis procedure (*Protocol 3*) they must be resuspended in 50 μl of methanol and then reacted with 50 μl of the freshly made o-PA solution. The samples should be incubated for at least 5 min in the dark as the reagent is light sensitive (note: samples should be kept in the dark throughout the HPLC run).

4. Add 0.2–0.5 ml of the running buffer to each tube.

5. Detection on HPLC is at an emission wavelength of 455 nm with excitation wavelength of 345 nm using a fluorescence detector.

6. Samples can now be injected on a reverse phase C18 column with a flow of 1 ml/min.

The quantitation of sphingoid backbones can either be done as a ratio to the internal standard and then normalized to phosphate, or in absolute amounts with the suggested external standard curve then normalized to phosphate. The case of using the ratio of experimental lipid over internal standard is done by simply dividing the peak area of experimental lipid by the peak area of the known standard. This gives a number that takes into consideration the extraction efficiency throughout the procedure and, once normalized to lipid phosphate, can be easily compared between control and treated samples. The reason for this is that the ratio does not change throughout experimental manipulations such as injections and quantitative transfers. Use of the external standard curve involves more manipulation. First, the peak area must be converted into picomoles via the standard curve generated. Next, one should determine the factors of how much sample was finally injected on the HPLC of the original third used. Therefore, one must account for extraction efficiency using the internal standard, amount injected over total amount, and any loss of sample through the quantitative transfer from the alkaline hydrolysis procedure. Once all these factors are accounted for one needs to calculate the true picomoles of each sphingoid backbone in each sample third and then normalize it to lipid phosphate. The absolute numbers can now be compared between control and treated samples.

5. Analysis of ceramides

Quantitation of ceramides and diacylglycerol (DAG) is accomplished most commonly by phosphorylation by the diacylglycerol kinase (DGK) method (9, 10). For the analysis of ceramides one should not use base hydrolysis because this procedure would hydrolyse the DAG. The basis of the assay is the ^{32}P-labelling of the ceramides using a tracer of hot γ-ATP which the kinase uses to phosphorylate the lipids. Once phosphorylated the lipids are re-extracted and then run out on thin layer chromatography (TLC) plates (*Figure 5*). Detection is done either by film or phosphoimager. Quantitation can be handled either by using Image Quant or by scraping the spots from the plate and doing scintillation counting. The protocol below uses Image Quant analysis to avoid the hazards of scraping (silica and dispersion of radiation).

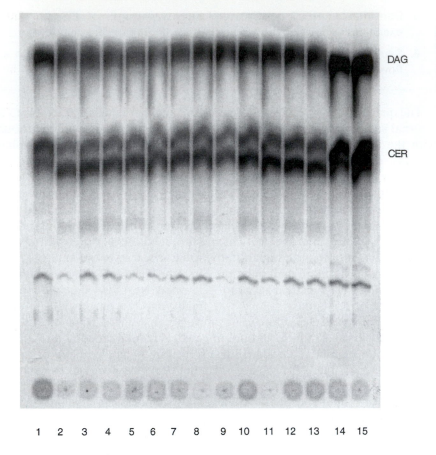

Figure 5. TLC of ceramide (CER) and diacylglycerol (DAG) post-DGK phosphorylation. Lanes are as follows: lane 1, 160 pmol standards of type III ceramide (Sigma) and diacylglycerol; lanes 2–13, MOLT4 samplés; lane 14, 320 pmol standards of type III ceramide and diacylglycerol; and lane 15, 640 pmol standards of type III ceramide and diacylglycerol.

Protocol 6. DGK of ceramides and DAG

Reagents
- ceramide (type III from Sigma)
- DAG
- β-octyl glucoside, βOG/dioleoylphos-phatidylglycerol, DOPG (7.5%:25 mM mixed micelle)
- 10 mM ATP

- chloroform/methanol (1:2)
- reaction mix: 103 mM imidazole, 17.9 MgCl$_2$, 2.86 mM dithiothreitol, 1.43 mM EGTA, 71.4 mM LiCl, 71.4 µg/ml DAG kinase, and 0.57 mM diethylenetriamine pentaacetic acid.

Method

1. Construct a standard curve of both ceramide and DAG, including a blank. The range of standards should be 40, 80, 160, 320, 640, and

116

1280 pmol. Combining the standards of ceramide and DAG into one tube per pmol amount helps to limit the assay size and the number of lanes to be spotted.

2. Dry down the standards and add to dried samples from either of the aforementioned extraction procedures (*Protocols 1* and *2*)

3. Resuspend the lipids with βOG/DOPG and vortex vigorously. If lipids do not go into solution, then sonicate and revortex.

4. Add 70 μl of reaction mixture to each sample and vortex well.

5. To start the reaction add 10 μl of ATP with an activity of 4.5 μCi per tube.

6. Allow reaction to proceed for 30 min at room temperature.

7. Stop the reaction with the addition of 3 ml of chloroform/methanol and vortex.

8. Add 1 ml of chloroform, 1.7 ml of water, and vortex to induce a phase break (Bligh and Dyer extraction conditions; *Protocol 1*).

9. Let samples sit briefly, then spin at 2000 *g* for 5 min to clarify phases.

10. Carefully aspirate off the upper phase (contains around 90% of the radiation) and collect 1.2–1.5 ml of the lower phase. Dry down the collected lipids via dry nitrogen or a speed vacuum.

11. Resuspend in 50 μl of chloroform and spot half on TLC plates (Whatman Silica 60A plates).

12. Develop plates in a TLC chamber containing the following solvent system: chloroform/acetone/methanol/acetic acid:water (10:4:3:2:1). Allow tank to equilibrate before running plates.

13. Expose plate to film overnight to obtain an image or directly go to an imager cassette with the phosphor screen and scan in using a phosphoimager. The resulting scan from the imager may be quantified using Image Quant.

The final quantitation of the ceramides and DAGs can be done after either counts per minute or band intensity have been determined. The actual picomoles should be obtained from the specific activity of ATP and, therefore, complete conversion of lipids by the kinase is required. Alternatively, quantitation of the mass may be obtained from the standard curve. Since the standard curve is internal and has been through all the experimental manipulations as have the samples, one can directly label them as their starting amounts. The standard curve is also a control for the assay in that it will show whether the assay is in the range of complete phosphorylation, and how well the assay was executed, by its linearity. Once conversion to picomoles has been done this number is for the whole third one started with and can be directly normalized to the third used for lipid phosphate. The minimal change in ceramide usually considered significant is a 50% increase or more and often a rise of twofold or more over basal.

6. Analysis of sphingomyelins

Presented in this section are two methods for the analysis of sphingomyelin. The first method is one of radiolabelling the sphingomyelin pool with tritiated choline, extracting the lipids, performing alkaline hydrolysis, and then measuring total sphingomyelin by counts released by incubation with bacterial sphingomyelinase (11) from *Streptomyces* (Sigma). The advantage of this method is the number of samples that may be analysed quickly; however, the method does not allow for analysis of particular sphingomyelins. The second method follows the same initial work-up as the first but then separates the sphingomyelins via HPLC (12) and fractions are collected and then counted. The advantage of this method is specific types of sphingomyelins (*Figure 6*) can be followed, but each sample takes a long time to run and count.

To quantitate the results one must calculate total cpms in each tube. Since we took 400 μl of the 800 μl upper phase (the upper phase is everything except chloroform), multiply the cpms by two. This value can now be

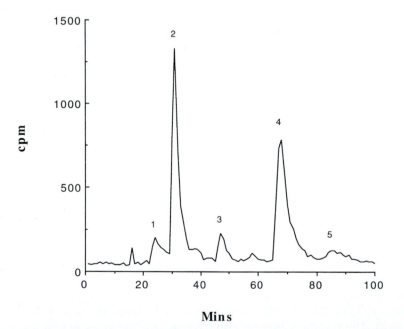

Figure 6. HPLC profiles of brain sphingomyelins. Peak 1 is 16:0 fatty acyl SM (24 min), peak 2 is 18:0 fatty acyl SM (31 min), peak 3 is 20:0 fatty acyl SM (47 min), peak 4 is a combination of 22:0 fatty acyl SM and 24:1 unsaturated fatty acyl SM (68 min), and peak 5 is 24:0 fatty acyl SM (85 min).

normalized to the phosphate value obtained above. Total sphingomyelin levels often decrease up to 30% with certain treatments.

Protocol 7. Bacterial sphingomyelinase on total SM

Reagents
- TMT buffer: 0.19 M Tris–HCl, pH 7.5, 12 mM MgCl$_2$, and 0.2% Triton X-100
- chloroform/methanol (2:1)
- bacterial SMase: 2 units/ml in 10 mM Tris–HCl, pH 7.5

Method
1. Suspend samples to be tested in 50 μl of TMT buffer and vortex vigorously.
2. Sonicate the samples for three one-minute bursts, vortex after each sonication then rest for 1 min between each sonication.
3. Pre-incubate the samples at 37 °C for 5 min.
4. Add 50 μl of bacterial Smase to each tube and allow the reaction to proceed for 30 min at 37 °C.
5. Stop the reaction with 1.5 ml of chloroform/methanol and vortex.
6. Add 0.2 ml of water to induce a phase break in the sample and vortex.
7. Spin tubes at 2000 *g* for 5 min to clarify phases.
8. Collect 400 μl of the upper phase for scintillation counting. (Remember, the cells were labelled with precursors to the polar head group and once incorporated will be released to the aqueous phase by enzyme cleavage.)

Protocol 8. HPLC on SM

Method
1. Prepare a known standard of radiolabelled SM for a control run.
2. Take samples post-extraction and alkaline hydrolysis and resuspend them in methanol.
3. Inject samples on to a C18 reverse phase column at a flow rate of 1 ml/min in a running system of 98% methanol, 2% 5 mM potassium phosphate, pH 7.0.
4. Count by either fraction collecting and scintillation counting or a flow through detector.

After counting each fraction a chart can be prepared for each sample. The charts of a treated versus non-treated sample can now be directly compared to see which specific SM species is changing. This comparison will allow analysis of each peak for changes in a specific sphingomyelin type.

7. Measurement of SMase and SM synthase activities

Further investigation of sphingolipid involvement in apoptosis can be gained by the analysis of the two key enzymes in the SM cycle. The most prominently measured enzymes are the sphingomyelinases, both acid and neutral. Basically, protein is extracted and incubated with radiolabelled substrate. The sphingomyelinase assay measures either an increase in enzyme levels and/or a post-translational modification activating or inactivating the enzyme, and not allosteric regulation *in vivo*. An enzyme of emerging importance is the SM synthase (13). The analysis of SM synthase activity is again accomplished by incubation under the proper conditions with radiolabelled ceramide. Again, the SM synthase assay can measure only increases in enzyme levels and/or post-translational modifications which regulate enzyme activity, and not other factors of control. These two enzyme assays can strengthen data on changes in the sphingolipids in a cell.

Protocol 9. Measurement of sphingomyelinase activity

Equipment and reagents

- PBS
- lysis buffer: 25 mM Tris–HCl, pH 7.4/5 mM EDTA, 20 μg/ml chymostatin, leupeptin, antipain and pepstatin, and 1 mM phenylmethylsulfonyl fluoride (PMSF)
- chloroform/methanol (2:1)
- assay buffer: 0.2M Tris–HCl, pH 7.4, 5mM MgCl$_2$, and 0.2% Triton X-100 for neutral SMase or 0.2M NaAC, pH 5.0, and 0.2% Triton X-100 for acid Smase
- nitrogen bomb

Method

1. Pellet 5×10^7 cells at 200 *g* in a table-top centrifuge at 4°C.
2. Resuspend in ice-cold PBS and then repellet as above.
3. Resuspend the cells in 0.5 ml lysis buffer.
4. Load cells into a nitrogen bomb on ice to perform nitrogen cavitation.
5. Assemble the nitrogen bomb, apply 350–500 psi of N$_2$ and let it sit on ice for 30 min.
6. Carefully open the outlet valve and allow cell homogenate to flow dropwise into a pre-chilled, 15 ml centrifuge tube.
7. Spin the homogenate at 1000 *g* for 10 min at 4°C.
8. Collect the supernatant (one can keep a crude homogenate for further analysis at this point).
9. Centrifuge the supernatant at 100 000 *g* for 1 h at 4°C.
10. Collect the supernatant from high speed spin (cytosol).

11. Resuspend the pellet in 0.5 ml of lysis buffer (membrane).

12. Determine protein concentrations via your favourite assay (e.g. Biorad).

13. Prepare the substrate. Each sample should have 4×10^5 dpm of labelled SM in 10 nmols of cold SM. Dry lipids down under nitrogen, then resuspend in appropriate amount of assay buffer. Vortex vigorously and sonicate if needed to completely solubilize the lipids.

14. Put 20–100 μg of cytosol or membrane protein into each tube. Use lysis buffer to bring the volume up to 100 μl. Set up blanks for each assay containing only 100 μl of lysis buffer.

15. To all tubes add 100 μl of SM substrate in the appropriate buffer.

16. Mix gently and incubate for 30 min at 37°C.

17. Add 1.5 ml of chloroform/methanol (2:1) to stop the reaction and vortex.

18. Add 0.4 ml of water to break the phases and vortex.

19. Centrifuge at 2000 g for 5 min in a table-top centrifuge to clarify the phases.

20. Take 400 μl of the upper phase and do scintillation counting. Remember to count 100 μl of substrate for total activity.

One has counted 400 of 900 μl, so one must determine total cpms hydrolysed from each sample. Now subtract the matching blank tube prepared for each assay. The number now obtained can be normalized to the amount of protein used in the assay.

Protocol 10. Measurement of sphingomyelin synthase activity

Equipment and reagents

- PBS
- lysis buffer: 25 mM Tris–HCl, pH 7.4, 5 mM EDTA, 1 mM phenylmethylsulfonyl fluoride, and 20 μg/ml of chymostatin, leupeptin, antipain, and pepstatin
- fatty acid-free BSA

- incubation buffer: 50 mM Tris–HCl, pH 7.4, 25 mM KCl, and 0.5 mM EDTA
- 20 nmoles [^{14}C]C$_6$ ceramide (9×10^3 cpm/nmole)
- 27 gauge 0.5 inch needle
- TLC plates

Method

1. Collect 1×10^7 cells via centrifugation at 500 g on a table-top centrifuge.

2. Wash cells with PBS and then repellet as above.

3. Suspend cells in approximately 0.8 ml of ice-cold lysis buffer.

4. Homogenize the cells by 10–15 passages through a 27 gauge 0.5 inch needle.

Protocol 10. *Continued*

5. Centrifuge homogenates for 10 min at 1000 *g* at 4°C to remove cellular debris and unbroken cells (One can keep part of this fraction as a crude homogenate).

6. Transfer the supernatant and spin at 100 000 *g* for 1 h at 4°C.

7. Collect the supernatant from high speed spin (cytosol).

8. Resuspend the pellet (membrane) in 0.4–0.5 ml of incubation buffer.

9. Determine protein concentrations via your favourite assay (e.g. Biorad).

10. Put 150 μg of protein from the above into a final volume of 0.5 ml incubation buffer. Set up blanks for each assay containing only 0.5 ml of incubation buffer.

11. Pre-incubate samples for 10 min at 37°C.

12. Start the reaction with the addition of $[^{14}C]C_6$ ceramide as an equimolar complex with fatty acid-free bovine serum albumin.

13. Allow the reaction to proceed for up to 15 min at 37°C.

14. Stop the reaction with 3 ml of chloroform/methanol (1:2), 0.2 ml of water, vortex, and keep on ice.

15. Add 1 ml of chloroform, 1 ml of water, and vortex. These additions induce a phase break. Allow the phase to clarify for 30 min, then centrifuge at 2000 *g* in a table-top centrifuge for 5 min to obtain clear phases.

16. One may now collect the lower organic phase directly or aspirate the aqueous phase and then collect the organic phase.

17. The extracted lipids should be in the 2 ml of chloroform and should be dried down via a speed vacuum apparatus or under a stream of dry nitrogen.

18. Resuspend in 50 μl of chloroform/methanol (1:2) and spot half on TLC plates (Whatman Silica 60A plates).

19. Take a small amount of the resuspended lipid and count to ensure similar cpms for each tube.

20. Develop the plates in a TLC chamber containing the following solvent system: chloroform/methanol/15 mM anhydrous $CaCl_2$ (60:35:8). Allow tank to equilibrate overnight before running plates.

21. The $[^{14}C]C_6$–SM band on the TLC plates is detected by autoradiography.

22. Scrape the band from the plate and do scintillation counting to yield cpms.

One has counted only half of the sample since this is what has been spotted. Subtract the matching blank tube prepared for each sample. Now multiply the obtained number by two as this will give you total cpms produced in each assay tube. The number now obtained can be normalized to the amount of protein used in the assay. Increased SM synthase activity should be reflected in a decrease of cellular ceramide levels.

Acknowledgements

This work was partly supported by NIH grant GM-43825.

References

1. Michel, C., van Echten-Deckert, G., Rother, J., Sandhoff, K., Wang, E., and Merrill, A. E. (1997). *Journal of Biological Chemistry*, **272**, 22432.
2. Perry, D., Obeid, L., and Hannun, Y. A. (1996). *Trends in Cardiovascular Medicine*, **6**, 158.
3. Obeid, L. M., Linardic, C. M., Karolak, L. A., and Hannun, Y. A. (1993). *Science*, **259**, 1769.
4. Kolesnick, R. and Fuks, Z. (1995). *Journal of Experimental Medicine*, **181**, 949.
5. Liu, B. and Hannun, Y. A. (1997). *Journal of Biological Chemistry*, **272**, 6281.
6. Bligh, E. G. and Dyer, W. J. (1959). *Canadian Journal of Biochemical Physiology*, **37**, 911.
7. Folch, J., Lees, M., and Sloane Stanley, G. H. (1957). *Journal of Biological Chemistry*, **226**, 497.
8. Merrill, A. H., Jr., Wang, E., Mullins, R. E., Jamison, W. C., Nimkar, S., and Liotta, D. C. (1988). *Analytical Biochemistry*, **171**, 373.
9. Preiss, J., Loomis, C. R., Bishop, W. R., Stein, R., Niedel, J. E., and Bell, R. M. (1986). *Journal of Biological Chemistry*, **261**, 8597.
10. Van Veldhoven, P. P., Bishop, W. R., and Bell, R. M. (1989). *Analytical Biochemistry*, **183** 177.
11. Jayadev, S., Linardic, C., and Hannun, Y. A. (1994). *Journal of Biological Chemistry*, **269** 5757.
12. Jungalwala, F. B., Hayssen, V., Pasquini, J. M., and McCluer, R. H. (1979). *Journal of Lipid Research*, **20**, 579.
13. Luberto, C. and Hannun, Y. A. (1998). *Journal of Biological Chemistry*, **73**, 14550.

7

Cytochemical detection of cytoskeletal and nucleoskeletal changes during apoptosis

MANON VAN ENGELAND, BERT SCHUTTE,
ANTON H. N. HOPMAN, FRANS C. S. RAMAEKERS, and
CHRIS P. M. REUTELINGSPERGER

1. Introduction

Apoptosis is an evolutionarily conserved form of physiological cell death by which redundant and damaged cells are eliminated during embryonic development and tissue homeostasis. In diseased tissue the process may be disturbed in such a way that accumulation of cells or increased cell loss may occur. Rapid elimination of the dying cell, without evoking an inflammatory response, is accomplished by disintegration of the cell into apoptotic bodies, which will then be phagocytosed (1). During this process, in which the plasma membrane remains intact, cleavage of cytoskeletal and nucleoskeletal proteins occurs by caspases, a family of cysteine proteases that cleave key proteins at conserved aspartic amino acid residues. The cleavage of cell–cell adhesion molecules leads to loss of contact with neighbouring cells (2), proteolysis of nuclear proteins facilitates nuclear disassembly (3), and the reorganization of actin filaments plays a role in the formation of apoptotic bodies (4). Proteolysis of fodrin may play a role in the loss of plasma membrane lipid asymmetry, resulting in exposure of phosphatidylserine (PS) at the outer leaflet of the cell membrane (5). Apoptosis can be detected on the basis of the characteristic morphological changes, such as chromatin aggregation and formation of apoptotic bodies. Furthermore, techniques have been developed which are based on apoptosis-specific biochemical changes (e.g. DNA fragmentation), expression of apoptosis-associated proteins, proteolysis of apoptosis-specific substrates, generation of apoptosis-specific neo-epitopes, or exposure of specific phospholipids at the outer plasma membrane. However, the individual techniques detect specific events in the process of apoptosis and therefore every technique has its limitations. In this chapter we describe methods to study changes in cyto- and nucleoskeletal proteins during apoptosis by means of immunocytochemistry. For detection of apoptotic cells the annexin V affinity assay and the TUNEL assay is used. The applications and limitations of the detection systems will be discussed.

2. Cyto- and nucleoskeletal changes during apoptosis

2.1 General

Apoptotic cells show characteristic morphological changes, such as loss of contact with neighbouring cells and with the substratum to which they adhere, nuclear condensation, and membrane blebbing. These morphological changes are, amongst others, the result of a dramatic reorganization of the cyto- and nucleoskeleton of the dying cell. Therefore, it is not surprising that many of the cyto- and nucleoskeletal components themselves are targets for the caspase family of proteases. However, in addition to caspase cleavage, post-translational modifications of these proteins, such as (de)phosphorylation, cross-linking, and citrullination are also reported to contribute to reorganization and modification of cyto- and nucleoskeletal proteins during apoptosis (6).

2.2 Microfilaments and microfilament-associated proteins

Reorganization of the microfilament network is involved in membrane blebbing during apoptosis. F-actin, present at the base of blebs during apoptosis (4, 7, 8), is necessary for bleb formation and the eventual formation of apoptotic bodies (9, 10). Whether actin itself is a direct target for caspases remains elusive. Several groups have proposed that actin is cleaved by caspases during apoptosis (11–14). However, Brown *et al.* (15) reported that in membrane-associated actin, a component of the cytoskeleton that links polymerized actin to the plasma membrane, cleavage occurred at a site devoid of a consensus motif for cleavage by caspases. Song *et al.* (16) reported protection of actin for proteolysis *in vivo*. Loss of contacts between neighbouring cells is caused by the disassembly of the cytoskeletal organization at the level of cell–cell adhesion sites. In fact, β-catenin, a known regulator of cell–cell adhesion, is proteolytically processed after induction of apoptosis. β-Catenin cleavage by caspase-3 removes the amino- and carboxy-terminal regions of the protein. The resulting β-catenin product is unable to bind α-catenin which is responsible for actin filament binding and organization (17). Also, cleavage of gelsolin (18) may be a physiological effector of morphological changes during apoptosis. The function of gelsolin is to bind barbed ends of actin monomers to prevent monomer exchange. Expression of the gelsolin cleavage product in multiple cell types caused the cells to round up, detach from the substratum, and undergo nuclear fragmentation (18).

2.3 Microtubules

Reorganization of tubulin is an integral part of the apoptotic process, independent of the involvement of tubulin in cell division (19). During apoptosis tubulin is reorganized into visible polymerized tubulin structures (20) which display characteristics similar to those observed following treatment with drugs that do interact directly with tubulin (21). Disruption of the micro-tubule architecture might influence signalling cascades. Recent studies by Blagosklonny *et al.* (22) have shown that disruption of microtubular archi-tecture leads to Raf-1 activation and Bcl-2 phosphorylation, serving a role similar to p53 induction following DNA damage.

2.4 Intermediate filaments

Cytokeratins, the intermediate filament proteins of epithelial cells, are necessary for the structural support of cells. During apoptosis, cytokeratin- and vimentin-containing intermediate filaments reorganize into granular structures (20, 23–25), as is illustrated in *Figure 1*. This process seems to be governed mainly by (hyper)phosphorylation of the cytokeratins during apoptosis (26).

Furthermore, it has been shown that phosphorylation and dephosphoryl-ation events also critically control cell junction assembly and stability and also regulate the formation of the cadherin–cytoskeleton complex (27, 28), thus influencing the adhesive properties of cells.

Recently Caulin *et al.* (24) and Leers *et al.* (25) showed that intermediate filament proteins are cleaved by caspases during apoptosis. A conserved caspase-6 motif is found in the linker-2 region of all intermediate filament proteins, except for the type II cytokeratins. Cytokeratin 18 has an additional caspase cleavage site at the C-terminus, as was suggested by Caulin *et al.* (24) and recently identified as the DALD motif by Leers *et al.* (25). The functional significance of the co-ordinated breakdown of the cytokeratins remains unknown. Caulin *et al.* (24) suggest that proteolytic cleavage of these inter-mediate filaments, which are highly polymerized and are relatively insoluble

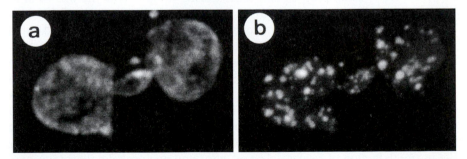

Figure 1. Linear projections of stacks of confocal scanning laser microscopy images of an apoptotic MR65 lung cancer cell after annexin V labelling, fixation, and staining with the M30 cytodeath antibody: (a) annexin V labelling, (b) aggregated cytokeratin 18 fragments. Apoptosis was induced by 6 h exposure of the cells to 50 μM roscovitine.

in normal aqueous solutions, is necessary for the formation of apoptotic bodies.

2.5 Nucleoskeletal components

Nuclear events in apoptosis include DNA condensation and fragmentation, redistribution of nuclear pores, disassembly of the nuclear lamina, and re-distribution and proteolysis of other nuclear proteins (29). Of the nuclear proteins proteolysed in apoptosis, several appear to be involved in DNA repair processes and in maintaining the proper conformation of chromatin through interactions with the nuclear matrix. Cleavage and reorganization of these proteins may contribute to the collapse of the nucleus in apoptosis. For example, proteolysis of lamin proteins facilitates nuclear disassembly (3). Lamins are intermediate filament proteins which form a lamina underlying the inner nuclear membrane. They have been shown to stabilize the nuclear envelope and to participate in determining the organization of the interphase nucleus (30). Two subtypes of lamins can be distinguished: i.e. the A-type lamins, which comprise lamin A, lamin AΔ10, and lamin C; and the B-type lamins, which comprise lamin B1 and lamin B2 (31). During apoptosis, lamins are cleaved by caspase-6 at a conserved VEID (A-type lamins) or VEVD (B-type lamins) amino acid sequence (3). Several of the nucleoskeleton proteins proteolysed in apoptosis appear to be autoantigens (32). Cleavage and modification of autoantigens may reveal immunocryptic epitopes that could potentially induce autoantibody response (32).

2.6 Generation of neo-epitopes in cytoskeletal proteins during apoptosis

Caspases are amongst the most specific endopeptidases known today. There-fore, activation of this family of proteases results in limited proteolytic cleavage at specific aspartate residues of the target molecules (33). A novel approach to probe for specific proteolytic events during apoptosis is the use of anti-bodies directed against neo-epitopes generated by caspase cleavage. This approach has been shown to be feasible for the caspase-cleaved cytoskeletal proteins actin and cytokeratin 18. Yang *et al.* (14) raised a polyclonal anti-serum using a synthetic peptide representing the last five amino acids of the C-terminus of the 32 kDa actin fragment produced during apoptosis. The antiserum specifically labels apoptotic but not necrotic cells. Similarly, Leers *et al.* (25) describe a monoclonal antibody specific for the liberated DALD sequence in the C-terminus of cytokeratin 18. Also this antibody specifically recognizes apoptotic cells and does not detect necrotic cells. Antibodies for the *in situ* detection of caspase-cleaved cyto- and nucleoskeletal proteins pro-vide simple, specific tools to evaluate the role of caspase activation in cyto-skeletal reorganization during apoptosis. They are now in use as apoptosis markers that can be applied to routinely processed tissues and cells.

3. Apoptosis detection systems

3.1 Overview

Apoptosis was originally detected because of the characteristic cell morphology, which allows apoptotic cells to be distinguished from healthy and necrotic cells. Although different cell types do not necessarily display all the hallmarks of apoptosis, there are similar features shared by cells undergoing apoptosis, which include shrinkage, blebbing, and chromatin condensation. Detection techniques based on apoptosis-specific biochemical changes, such as DNA fragmentation, exposure of phosphatidylserine (PS) at the outer plasma membrane, and proteolysis of proteins have been developed. In the next paragraphs, we describe protocols to study cyto-and nucleoskeletal proteins during apoptosis using two different apoptosis detection systems, i.e. the annexin V affinity assay to detect PS exposure and the TUNEL assay to detect DNA fragmentation.

3.2 The annexin V affinity assay

In healthy cells, the phospholipids of the plasma membrane are distributed asymmetrically over the two leaflets of the bilayer. Phosphatidylcholine and sphingomyelin are the predominant species of the outer membrane leaflet and PS is located exclusively in the inner membrane leaflet. This asymmetry is maintained by an aminophospholipid translocase which transports the aminophospholipids, with preference for PS, from the outer to the inner leaflet (34). In addition, the localization of PS to the inner membrane leaflet is maintained by association of PS with annexins, polyamines, and membrane skeletal proteins such as fodrin (35).

Early in apoptosis, PS is externalized to the outer membrane leaflet due to inhibition of the aminophospholipid translocase and subsequent activation of a scramblase (36, 37). Also, fodrin cleavage by caspase-3 and calpain may contribute to PS externalization (38).

Exposure of PS at the outer surface of the plasma membrane serves as a trigger for macrophages, which posses a PS receptor on their plasma membrane, to engulf and digest the apoptotic cell. Exposure of PS on the plasma membrane of the apoptotic cell seems to be a phylogenetically conserved phenomenon which occurs in mammalian, avian, insect, and plant cells (39–41) and can be detected *in vitro* and *in vivo* using annexin V (42). Annexin V was first discovered by Bohn and co-workers (43) who isolated the protein from human placenta and called it placental protein (PP4). Later, the same protein was isolated from human umbilical cord by Reutelingsperger *et al.* (44) who called it vascular anticoagulant protein α (VACα). Meanwhile, other groups also isolated the protein and referred to it as inhibitor of blood coagulation (IBC), endonexin II, placental anticoagulant protein I (PAP-I), lipocortin V, 35 K calelectrin, and anchorin II (45). The function of annexin V *in vivo* is still unknown, but from *in vitro* experiments it is has become clear that it has anti-phospholipase, anti-coagulant, anti-kinase, and phospholipid-binding activities (46). Annexin V binds PS in the presence of calcium ions. Therefore, the availability of biotin- or fluorochrome-labelled annexin V enables one to study PS exposure in cells undergoing apoptosis.

Protocol 1a. *In vitro* detection of apoptotic cells by annexin V affinity labelling. Cells grown in suspension analysed by flow cytometry or fluorescence microscopy

Equipment and reagents

- cell line of interest
- apoptosis-inducing agent of interest
- annexin V-binding buffer (25 mM Hepes, 125 mM NaCl, 2.5 mM CaCl$_2$, included in APOPTEST™ kit)[a]
- flow cytometer or fluorescence microscope

- fluorochrome-conjugated annexin V: APOPTEST™–FITC, APOPTEST™–oregon green, APOPTEST™–phycoerythrin[b]
- propidium iodide [PI, included in APOPTEST™ kit, stock solution of 250 μg/ml in PBS/RNAse A (10 mg/ml)]

Method

1. Add fluorochrome-labelled annexin V (dilute according to the manufacturer's instructions) to the apoptotic cells (10^6 cells/ml) in culture medium or in annexin V-binding buffer.[c, d]

2. Incubate for 3–5 min at room temperature.

3. Add PI to the cell suspension (final concentration 5 μg/ml for flow cytometry and 1 μg/ml for fluorescence microscopy) and incubate on ice for 15 min.[e,f]

4. Quantify annexin V binding by flow cytometry or place a drop of cells on a glass slide, add a coverslip, and visualize the cells using fluorescence microscopy.

[a] All APOPTEST™ products are obtained from Nexins Research.
[b] Annexin V–phycoerythrin can only be used for flow cytometry.
[c] For double-labelling with PI, annexin V–oregon green or annexin V–FITC is recommended.
[d] It is essential that the culture medium and the annexin V-binding buffer contain 2.5 mM calcium, since binding of annexin V to PS is calcium dependent.
[e] Annexin V is able to detect PS at the outer surface of the plasma membrane. However, annexin V also binds PS present at the cytoplasmic site of the cell membrane. Thus it will bind necrotic cells of which the plasma membrane is permeable. Therefore, in *in vitro* assays, a membrane-impermeable vital dye, such as PI, must be included in the assay to monitor plasma membrane integrity and to discriminate between apoptotic and necrotic cells.
[f] Using the annexin V/PI assay in viable cells, three populations of cells can be detected: (a) viable cells which are annexin V negative/PI negative; (b) apoptotic cells which are annexin V positive/PI negative; and (c) (secondary) necrotic cells which are annexin V positive/PI positive.

Koopman *et al.* (47) were the first to describe detection of apoptotic cells *in vitro* using FITC-labelled annexin V. Hapten-labelled annexin V can bind to externalized PS in the outer plasma membrane leaflet of apoptotic cells. It will not bind to normal cells because the molecule cannot trespass the phospholipid bilayer. In necrotic cells, however, the inner leaflet of the plasma membrane is available for binding, since the integrity of the plasma membrane is lost. To discriminate between necrotic and apoptotic cells, a membrane-impermeable DNA dye, for example propidium iodide (PI), can be included in the assay. In this way vital, apoptotic, and necrotic cells can be discriminated on the basis of double-labelling for annexin V and PI and analysed by flow cytometry and fluorescence microscopy.

For cells cultured in suspension, detection of PS exposure is relatively easy (*Protocol 1a*). For adherent cells, however, the situation is more complex. When cells are grown on glass slides, they can be analysed immediately after annexin V labelling using a fluorescence microscope. To obtain a single-cell suspension of adherent cells for flow cytometric analysis, cells have to be harvested. However, enzymatic harvesting of adherent cell lines may interfere with reliable detection of PS exposure and trypsin or ethylenediaminotetra-acetic acid (EDTA) harvesting of cells before annexin V labelling can induce changes in the plasma membrane, which leads to false-positive results. Trypsin or EDTA treatment after annexin V labelling interferes with the detection of bound annexin V, because trypsin and EDTA remove bound annexin V by proteolysis and chelation of calcium ions, respectively. Therefore, a method is developed (*Protocol 1b*) to detect PS exposure in adherent cell lines by labelling cells as a monolayer with annexin V and by harvesting the cells thereafter by means of scraping (48).

Protocol 1b. *In vitro* detection of apoptotic cells by annexin V affinity labelling. Adherent cells analysed by flow cytometry or fluorescence microscopy

Equipment and reagents
- adherent cell line of interest, grown in a culture flask or on glass slides
- apoptosis-inducing agent of interest
- annexin V-binding buffer (see *Protocol 1a*)
- fluorochrome-conjugated annexin V (see *Protocol 1a*)
- rubber policeman
- PBS/PI/RNAse (see *Protocol 1a*)

Method
1. Add fluorochrome-labelled annexin V (dilute according to the manufacturer's instructions) to the adherent cells in the culture flask or the cells grown on slides without washing the cells.[a-c]
2. Incubate for 3–5 min in the culture flask or on glass slides. Cells labelled with annexin V–FITC or annexin V–oregon green can now be visualized immediately after addition of PI (see step 7).
3. Collect the culture medium (this fraction contains the detached apoptotic cells).

Protocol 1b. *Continued*

4. Wash the adherent cells three times with culture medium or annexin V-binding buffer to remove excess annexin V, collect the rinse solution, and pool with the collected culture medium from step 3. Pellet these cells by centrifugation (400 *g*, 5 min, 4°C), discard the supernatant, resuspend the pellet in culture medium or annexin V buffer, pellet the cells again by centrifugation, discard the rinse medium, and repeat this washing procedure.[d]

5. Detach the adherent cells in culture medium gently by scraping with a rubber policeman, pellet the cells by centrifugation, and discard the culture medium.[e]

6. Resuspend the pellet of detached cells in culture medium or annexin V buffer and pool these cells with the cells collected in step 4. Dilute to a final concentration of 10^6 cells/ml.

7. Add PI to the cell suspension (final concentration 5 μg/ml for flow cytometry and 1 μg/ml for fluorescence microscopy) and incubate on ice for 15 min.

8. Analyse the cells by flow cytometry or visualize cells using fluorescence microscopy.

[a] Annexin V–phycoerythrin can only be used for flow cytometry.
[b] For double-labelling with PI, annexin V–oregon green or annexin V–FITC is recommended.
[c] It is essential that the culture medium and the annexin V-binding buffer contain 2.5 mM calcium since binding of annexin V to PS is calcium dependent. The washing step is omitted to obtain an optimal number of apoptotic cells, which would otherwise be rinsed away.
[d] Removal of annexin V before scraping of the adherent cells prevents annexin V binding to internally located PS in cells that become damaged (permeable) during the scraping procedure.
[e] To avoid false-positive and false-negative results due to trypsin or EDTA treatment, annexin V-labelled cells are harvested using a rubber policeman. This method of harvesting introduces a fourth population of cells, i.e. damaged cells, which are annexin V negative and PI positive.

To study PS exposure and cyto- and nucleoskeletal reorganizations in apoptosis simultaneously, using annexin V and immunocytochemistry, cells have to be permeabilized for antibodies to enter the cell and to detect their epitopes. Therefore a method (*Protocol 2*) is developed in which the cells are fixed after labelling with annexin V (20).

In some circumstances the annexin V affinity assay cannot be used to detect apoptotic cells. In frozen and paraffin-embedded tissue sections, for example, annexin V also detects PS at the inner plasma membrane leaflet. Therefore, another apoptosis detection system has to be chosen to study cyto- and nucleoskeletal changes in tissue sections. In tissue sections the TUNEL (49) and (50) DNA end-labelling procedures are often used to detect apoptotic cells. The methods described below can be applied to tissue sections as well as cells in culture.

Protocol 2. Combined annexin V labelling and nucleo- and cytoskeletal protein immunostaining of cells analysed by flow cytometry or fluorescence microscopy

Equipment and reagents

- biotin-conjugated annexin V (APOPTEST™–biotin) or oregon green-conjugated annexin V (APOPTEST™–oregon green)
- binding buffer (see *Protocol 1a*)
- primary antibody directed against the antigen of interest, e.g. mouse monoclonal antibody directed against actin or cytokeratin

- PBS
- PBS/BSA (1%, Sigma)
- fluorochrome-conjugated secondary antibody, e.g. phycoerythrin- or Texas Red-conjugated rabbit anti-mouse antibody
- PBS/PI/RNAse (see *Protocol 1a*) or PBS/4',6-diamino-2-phenyl indole (DAPI, 0.5 µg/ml)

Method

1. Depending on the cell line, use *Protocol 1a* or *Protocol 1b* (steps 1–2 and 1–6, respectively) to label cells with annexin V. In this case, use annexin V–oregon green or annexin V–biotin.[a, b]

2. In the case of cells in suspension or adherent cells harvested for flow cytometry, pellet the cells by centrifugation (400 *g*, 5 min, 4°C), resuspend the cells in binding buffer, pellet again, and discard the binding buffer.[c]

3. Fix the cells in methanol for 5 min at –20°C.[d]

4. Pellet the cells by centrifugation, discard the methanol, and resuspend the cells in PBS.

5. Pellet the cells by centrifugation, discard the PBS, and resuspend the cells in PBS/BSA (1%).

6. Incubate the cells with the (appropriately diluted) primary antibody for 60 min at room temperature.

7. Pellet the cells by centrifugation, discard the PBS/BSA, and resuspend the cells in PBS/BSA, repeat once more.

8. Incubate the cells with the (appropriately diluted) secondary antibody and in the case of annexin V–biotin labelling, incubate with fluorochrome-conjugated avidin (e.g. FITC-conjugated avidin) for 60 min at room temperature.[e, f]

9. Pellet the cells, discard the PBS/BSA, and resuspend the cells in PBS/BSA, repeat once.

10. Prepare a cell suspension in PBS/BSA of 10^6 cells/ml.

11. For flow cytometry: counterstain cells with PI (final concentration 1 µg/ml) and incubate on ice for 15 min.[g, h] For fluorescence microscopy: counterstain cells with PBS/DAPI.

Protocol 2. *Continued*

12. Analyse cells by flow cytometry (see *Figure 2*) or place a drop of cells on a glass slide, add a coverslip, and visualize the cells by fluorescence microscopy (see *Figure 1*).

---

[a]The fluorescent properties of FITC are impaired by methanol fixation, therefore the use of annexin V–oregon green (low photobleaching and pH insensitivity) or annexin V–biotin is recommended.

[b]APOPTEST™–biotin, product B700 is recommended in *in vitro* experiments.

[c]It is possible to stain adherent cells grown on glass slides simultaneously for PS exposure and cyto-and nucleoskeletal proteins. However, it is recommended to harvest adherent cells (*Protocol 1b*, steps 3–6) after annexin V labelling, because fixation and repeated washing procedures during immunocytochemistry often lead to detachment and subsequent loss of apoptotic cells.

[d]Be aware of the fact that cyto- and nucleoskeletal proteins (or fragments) can be lost from apoptotic cells using methanol fixation due to increased solubility of the protein or fragments of proteins.

[e]Use two different fluorochromes to detect annexin V and antibody binding.

[f]A phycoerythrin-conjugated secondary antibody is recommended for flow cytometry.

[g]After fixation of cells, PI cannot be used to discriminate between apoptotic and necrotic cells by membrane permeability. In these cells PI indicates DNA content of the cells.

[h]The concentration PI has to be decreased in a multiparameter flow cytometric analysis compared with a single colour analysis because of the spectral overlap between the red colours phycoerythrin and PI.

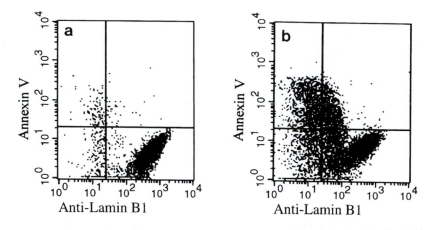

Figure 2. Flow cytograms of (a) control and (b) apoptotic MR65 lung cancer cells after annexin V labelling, fixation, and staining with a lamin B1 antibody. Quadrant settings were based on the negative control. (a) In the control culture only a few annexin V positive cells (spontaneously apoptotic cells) are observed. The majority of the cells are positive for lamin B1 and negative for annexin V. (b) After induction of apoptosis with roscovitine, part of the cells becomes annexin V positive. The lamin immunofluorescence of these cells is decreased.

3.3 Detection of apoptosis using the TUNEL assay

DNA fragmentation is considered to be a key event in apoptosis and can be detected as a typical DNA ladder on agarose gels. However, this method does not provide information regarding apoptosis in individual cells and cannot be used to evaluate apoptosis in relation to histological localization. This can be achieved by enzymatic *in situ* labelling of 3'-OH ends of fragmented DNA. For incorporation of labelled nucleotides into DNA strand breaks, terminal deoxynucleotidyl transferase (TdT) can be used. This method is known as the TUNEL (TdT-mediated conjugated dUTP nick end-labelling) assay (49). If DNA polymerase I is used to incorporate the nucleotides the method is known as the ISNT (*in situ* nick translation) assay (50). The disadvantage of these methods is that late necrotic cells, showing DNA degradation, will also be labelled. In *Protocol 3* a method is described to combine the TUNEL assay and immunostaining of cyto-and nucleoskeletal proteins in cell lines, while in *Protocol 4* a method is described to combine the TUNEL assay and the immunocytochemical staining of cyto-and nucleoskeletal proteins in paraffin-embedded tissue.

Protocol 3. Combined detection of DNA fragmentation (TUNEL) and cyto- and nucleoskeletal changes in apoptotic cells using flow cytometry or fluorescence microscopy of cell lines

Equipment and reagents

- cell line of interest in suspension (for flow cytometry) or cytocentrifuged on to glass slides (for fluorescence microscopy)[a]
- *in situ* cell death detection kit, fluorescein (Boehringer Mannheim), including TdT enzyme, FITC-labelled 11-deoxyuridine triphosphate (dUTP–FITC), and TdT reaction buffer containing nucleotide mixture.
- primary antibody directed against the antigen of interest (e.g. a mouse monoclonal antibody raised against lamin or cytokeratin)

- fluorochrome-labelled secondary antibody (e.g. Texas Red or phycoerythrin-conjugated rabbit anti-mouse antibody)[b]
- $4 \times$ SSC
- PBS/BSA (1%, Sigma)
- PBS/PI/RNAse [stock solution of 250 µg/ml PI in PBS/RNAse (10 mg/ml)] for flow cytometry or PBS/4',6-diamino-2-phenyl indole (DAPI, final concentration 0.5 µg/ml) for immunocytochemistry
- flow cytometer or fluorescence microscope

Method

1. Fix the cells or slides in cold methanol (–20°C) for 5 min.

2. Rinse twice with PBS.

3. Prepare the TUNEL reaction mix according to the manufacturers instructions.

4. Incubate with the TUNEL reaction mix for 60 min at 37°C in a humidified atmosphere in the dark. In this step, incorporation of FITC–dUTP to 3'-OH DNA fragments by the enzyme TdT is achieved.

Protocol 3. *Continued*

5. Rinse in 4 × SSC.

6. Rinse twice in PBS/BSA.

7. Incubate cells with the primary antibody for 60 min at room temperature in the dark.

8. Rinse twice with PBS/BSA.

9. Incubate with a fluorochrome (non-FITC)-conjugated secondary antibody to visualize the antibody of interest for 60 min at room temperature in the dark.[c]

10. Rinse twice with PBS/BSA.

11. For flow cytometry: add PI to the cells (final concentration 1 µg/ml) and incubate on ice for 15 min.[d] For fluorescence microscopy: add DAPI to the cells (final concentration 0.5 µg/ml).

12. Analyse cells by flow cytometry (see *Figure 3*) or fluorescence microscopy.

[a] It is recommended to harvest adherent cells (*Protocol 1b*, steps 3–6) after annexin V labelling, because fixation and repeated washing procedures during incorporation of dUTP and immunocytochemistry often lead to detachment and subsequent loss of apoptotic cells from the slide.
[b] For flow cytometry, a phycoerythrin-conjugated secondary antibody is recommended. For fluorescence microscopy, a Texas Red-conjugated secondary antibody is recommended.
[c] Because the dUTP is conjugated with FITC, use a non-FITC-conjugated secondary antibody.
[d] The concentration of PI has to be decreased in a multiparameter flow cytometric analysis compared with a single colour analysis because of the overlap between the red colours.

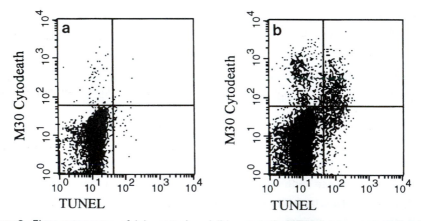

Figure 3. Flow cytograms of (a) control and (b) apoptotic MR65 lung cancer cells after TUNEL reaction and staining of cells with the M30 cytodeath antibody. Quadrant settings were based on the negative control. (a) In the control culture the majority of the cells are TUNEL negative and M30 negative. (b) After induction of apoptosis a fraction of the cells has become M30 positive, indicating cleavage of cytokeratin 18 by caspases. The M30 positive cells are partly positive and partly negative for TUNEL reaction.

The use of fluorescence microscopy on paraffin-embedded tissue is often hampered by a high autofluorescence background present in the tissue section. Therefore a protocol (*Protocol 4*) is described to study DNA fragmentation and cyto- and nucleoskeletal changes simultaneously using brightfield microscopy.

Protocol 4. Combined detection of DNA fragmentation (TUNEL) and cyto-and nucleoskeletal changes in apoptotic cells using brightfield microscopy on paraffin-embedded tissue

Equipment and reagents

- 3–5 μm thick paraffin-embedded tissue sections mounted on to glass slides
- xylene
- ethanol series (absolute, 95%, 90%, 80%, 70%, diluted in distilled water)
- 0.3% H_2O_2/methanol
- proteinase K (Sigma, final concentration 20 μg/ml in 10 mM Tris–HCl, pH 7.4–8.0)
- *in situ* cell death detection kit, alkaline phosphatase (Boehringer Mannheim), including TdT enzyme, FITC-labelled 11-deoxyuridine triphosphate (dUTP–FITC), TdT reaction buffer containing nucleotide mixture, alkaline phosphatase-conjugated sheep anti-FITC antibody
- PBS
- PBS/BSA (1%)

- chromogenic substrate for alkaline phosphatase reaction (e.g. N-ASMX-phosphate/new fuchsin)
- primary antibody directed to the antigen of interest, e.g. monoclonal mouse antibody M30 cytodeath (Boehringer Mannheim), directed against caspase-cleaved cytokeratin 18 fragment
- peroxidase-conjugated secondary antibody, e.g. peroxidase-conjugated rabbit anti-mouse antibody
- chromogenic substrate for peroxidase reaction [e.g. diaminobenzidine tetrachloride (DAB), Sigma]
- haematoxylin
- mounting medium for light microscopy (e.g. Aquamount, BDH)

Method

1. Deparaffinize and rehydrate tissue section using xylene and descending ethanol series.

2. Rinse slides in PBS.

3. Block endogenous peroxidase activity by incubating slides with 0.3% H_2O_2/methanol.

4. Rinse slides in PBS.

5. Incubate slides with proteinase K (final concentration 20 μg/ml) for 15 min at 37°C.[a]

6. Rinse slides in PBS.

7. Prepare the TUNEL reaction mix according to the manufacturer's instructions.

8. Incubate slides with 50 μl TUNEL reaction mix for 60 min at 37°C in a humidified chamber. In this step, incorporation of FITC–dUTP to 3'-OH DNA fragments by the enzyme TdT is achieved.[b]

9. Rinse slides with 4 × SSC buffer.

Protocol 4. *Continued*

10. Rinse slides with PBS.

11. Incubate slides with 50 µl alkaline phosphatase-labelled converter antibody directed against FITC for 30 min at 37°C in a humidified chamber.

12. Rinse slides three times with PBS.

13. Detect alkaline phosphatase activity by precipitating chromogenic substrates, e.g. N-ASMX-phosphate/new fuchsin.[c]

14. Rinse slides in distilled water.

15. Rinse slides in PBS/BSA.

16. Incubate with the primary antibody for 60 min at room temperature in a humidified chamber.

17. Rinse slides three times with PBS.

18. Incubate with the peroxidase-conjugated secondary antibody for 60 min at room temperature in a humidified chamber.

19. Rinse three times with PBS.

20. Detect peroxidase activity by precipitating chromogenic substrates, e.g. DAB.

21. Rinse slides in distilled water.

22. Counterstain slides with hematoxyline.

23. Embed slides in an aqueous mounting medium for light microscopy.

24. Analyse slides with a brightfield microscope.

[a] Concentration, time, and temperature of the pre-treatment steps have to be optimized for each type of tissue and for each type of antibody. The described conditions are the conditions optimized for the M30 antibody, directed against a caspase-cleaved cytokeratin 18 fragment. Use of other antibodies directed against cyto-and nucleoskeletal proteins may need specific adaptations of the protocol concerning epitope recognition.

[b] At this stage samples can be visualized under a fluorescence microscope.

[c] For more information on chromogenic substrates that can be combined, see Speel *et al.* (51).

References

1. Wyllie, A.H. (1993). *Br. J. Cancer*, **67**, 205.
2. Herren, B., Levkau, B., Raines, E.W. and Ross, R. (1998). *Mol. Biol. Cell*, **6**, 1589.
3. Rao, L., Perez, D. and White, E. (1996). *J. Cell Biol.*, **135**, 1441.
4. Laster, S.M. and MacKenzie, J.M. (1996). *Microsc. Res. Tech.*, **34**, 272.
5. Martin S.J., Reutelingsperger, C.P.M., McGahon, A.J., Rader, J.A., van Schie, R.C., LaFace, D.M. and Green, D.R. (1995). *J. Exp. Med.*, **182**, 1545.
6. Utz, P.J. and Anderson, P. (1998). *Arthritis Rheum.*, **41**, 1152.
7. Pitzer, F., Dantes, A., Fuchs, T., Baumeister, W. and Amsterdam, A. (1996). *FEBS Lett.*, **394**, 47.

8. Vemuri, G.S., Zhang, J., Huang, R., Keen, J.H. and Rittenhouse, S.E. (1996). *Biochem. J.*, **314**, 805.

9. Cotter, T.G., Lennon, S.V., Glynn, J.M. and Green, D.R. (1992). *Cancer Res.*, **52**, 997.

10. Levee, M.G., Dabrowska, M.I., Lelli, J.L. and Hinshaw, D.B. (1996). *Am. J. Physiol.*, **271**, C1981.

11. Mashima, T., Naito, M., Fujita, N., Noguchi, K. and Tsuruo, T. (1995). *Biochem. Biophys. Res. Commun.*, **217**, 1185.

12. Kayalar, C., Ord, T., Testa, M. P., Zhong, L.T. and Bredesen, D.E. (1996). *Proc. Natl. Acad. Sci. USA*, **93**, 2234.

13. McCarthy, N.J., Whyte, M.K.B., Gilbert, C.S. and Evan, G.I. (1997). *J. Cell Biol.*, **136**, 215.

14. Yang, F., Sun, X., Beech, W., Teter, B., Wu, S., Sigel, J., Vinters, H.V., Frautschy, S.A. and Cole, G.M. (1998). *Am. J. Pathol.*, **152**, 379.

15. Brown, S.B., Bailey, K. and Savill, J. (1997). *Biochem. J.*, **323**, 233.

16. Song, Q., Wei, T., Lees-Miller, S., Alnemri, E., Watters, D. and Lavin, M.F. (1997). *Proc. Natl. Acad. Sci. USA*, **94**, 157.

17. Brancolini, C., Lazarevic, D., Rodriguez, J. and Schneider, C. (1997). *J. Cell. Biol.*, **139**, 759.

18. Kothakota, S., Azuma, T., Reinhard, C., Klippel, A., Tang, J., Chu, K., McGarry, T.J., Kirschner, M.W., Koths, K., Kwiatkowski, D.J. and Williams, L.T. (1997). *Science*, **278**, 294.

19. Pittman, S.M., Strickland, D. and Ireland, C.M. (1994). *Exp. Cell Res.*, **215**, 263.

20. van Engeland, M., Kuijpers, H.J.H., Ramaekers, F.C.S., Reutelingsperger, C.P.M. and Schutte, B. (1997). *Exp. Cell Res.*, **235**, 421.

21. Ireland, C.M. and Pittman, S.M. (1995). *Biochem. Pharmacol.*, **49**, 1491.

22. Blagosklonny, M.V., Giannakakou, P., el-Deiry, W.S., Kingston, D.G., Higgs, P.I., Neckers, L.and Fojo, T. (1997). *Cancer Res.*, **57**, 130.

23. Tinnemans, M.M.F.J., Lenders, M.H.J.H., ten Velde, G.P.M., Ramaekers, F.C.S. and Schutte, B. (1995). *Eur. J. Cell Biol.*, **68**, 35.

24. Caulin, C., Salvesen, G.S. and Oshima, R.G. (1997). *J. Cell Biol.*, **138**, 1379.

25. Leers, M.P.G., Kölgen, W., Björklund, V., Bergman, T., Tribbick, G., Persson, B., Björklund, P., Ramaekers, F.C.S., Björklund, B., Nap, M., Jörnvall, H. and Schutte, B. (1999). *J. Pathol.*, **187**, 567.

26. Liao, J., Ku, N.O. and Omary, M.B. (1997). *J. Biol. Chem.*, **272**, 17565.

27. Serres, M., Grangeasse, C., Haftek, M., Durocher, Y., Duclos, B. and Schmitt, D. (1997). *Exp. Cell Res.*, **231**, 163.

28. Aberle, H., Bauer, A., Stappert, J., Kispert, A. and Kemler, R. (1997). *EMBO J.*, **16**, 3797.

29. Earnshaw, W.C. (1995). *Curr. Opin. Cell Biol.*, **7**, 337.

30. Moir, R.D. and Goldman R.D. (1993). *Curr. Opin. Cell. Biol.*, **5**, 408.

31. Moir R.D., Spann. T.P. Goldman, R.D. (1995). *Int. Rev.Cytol.*, **162B**, 141.

32. Casiano, C.A., Martin, S.J., Green, D.R. and Tan, E.M. (1996). *J. Exp. Med.*, **184**, 765.

33. Salvesen, G.S. and Dixit, V.M. (1997). *Cell*, **91**, 443.

34. Diaz, C. and Schroit, A.J. (1996). *J. Membr. Biol.*, **151**, 1.

35. Fadok, V.A., Bratton, D.L., Frasch, S.C., Warner, M.L. and Henson, P.M. (1998). *Cell Death Differ.*, **5**, 551.

36. Zwaal, R.F.A. and Schroit, A.J. (1997). *Blood*, **89**, 1121.
37. Verhoven, B., Schlegel, R.A., Williamson, P. (1995). *J. Exp. Med.*, **182**, 1597.
38. Vanags D.M., Porn Ares, M.I., Coppola, S., Burgess, D.H. and Orrenius, S. (1996). *J. Biol. Chem.*, **271**, 31075.
39. Fadok, V.A., Voelker, D.R., Campbell, P.A., Cohen, J.J., Bratton, D.L. and Henson, P.M. (1992). *J. Immunol.*, **148**, 2207.
40. van den Eijnde, S.M., Boshart, L., Reutelingsperger, C.P.M., de Zeeuw, C.I., Vermeij-Keers, C. (1997). *Cell Death Differ.*, **4**, 311.
41. O'Brien, I.E.W., Reutelingsperger, C.P.M. and Holdaway, K.M. (1997). *Cytometry*, **29**, 28.
42. van Engeland, M., Nieland, L.J.W., Ramaekers, F.C.S., Schutte, B. and Reutelingsperger C.P.M. (1998). *Cytometry*, **31**, 1.
43. Inaba, N., Sato, N., Ijichi, M., Fukazawa, I., Nito, A., Takamizawa, H., Luben, G., Bohn, H. (1984). *Tumour Biol.*, **5**, 75.
44. Reutelingsperger, C.P.M, Hornstra, G. and Hemker, H.C. (1985). *Eur. J. Biochem.*, **151**, 625.
45. van Heerde, W.L., Degroot, P.G. and Reutelingsperger, C.P.M. (1995). *Thromb. Haemost.*, **73**, 172.
46. Reutelingsperger, C.P.M. and van Heerde, W.L. (1997). *Cell. Mol. Life Sci.*, **53**, 527.
47. Koopman, G., Reutelingsperger, C.P.M., Kuijten, G.A.M., Keehnen, R.M.J., Pals, S.T. and van Oers, M.H.J. (1994). *Blood*, **84**, 1415.
48. van Engeland, M., Ramaekers, F.C.S., Schutte, B. and Reutelingsperger, C.P.M. (1996). *Cytometry*, **24**, 131.
49. Gavrieli, Y., Sherman, Y. and Ben-Sasson, S.A. (1992). *J. Cell. Biol.*, **119**, 493.
50. Wijsman, J.H., Jonker R.R., Keijzer, R., van de Velde, C.J.H., Cornelisse, C.J. van Dierendonck, J.H. (1993). *J. Histochem. Cytochem.*, **41**, 7
51. Speel, E.J.M., Ramaekers, F.C.S and Hopman, A.H.N. (1995). *Histochem. J.*, **27**, 833.

8

Metabolic alterations associated with apoptosis

ANA P. COSTA-PEREIRA and THOMAS G. COTTER

1. Introduction

Apoptosis is an active form of cell death, induced both by endogenous and exogenous stimuli, which occurs under physiological conditions (1, 2). Morphologically, it is characterized by cellular features such as membrane blebbing, cell shrinkage, and chromatin condensation (3). Intracellular events include various types of DNA fragmentation (4), cytoskeletal alterations, and protease activation (reviewed in ref. 5).

Oxidative stress is believed to play a key role in the process of apoptosis. Indeed, accumulation of oxidized proteins and lipids (6), rapid production of reactive oxygen intermediates (ROI), and alterations of the cellular redox (7, 8), as well as disruption of the transmembrane potential ($\Delta\psi_m$) (9), have all been reported to be common metabolic alterations during the apoptosis of a variety of cell types. Oxidative stress is also seen in a number of pathological conditions (reviewed in ref. 10). The degeneration of neural cells observed during the progression of Alzheimer's, Parkinson's, and other neuro-degenerative diseases have been linked to the cumulative damage caused by oxidative stress. The levels of glutathione (GSH), a major component of the intracellular antioxidant defence, have been shown to be decreased in HIV-infected cells (11). In addition, *in vitro* studies on cells from HIV patients have shown the decrease of several other antioxidant enzymes (12). Moreover, tumour promotion during oncogenesis is favoured under pro-oxidant conditions and several antineoplastic drugs have been demonstrated to be potent antioxidants (13).

This chapter describes some current procedures used to analyse common intracellular metabolic alterations which are of importance in apoptosis, namely the disruption of $\Delta\psi_m$, generation of ROI (e.g. peroxide and superoxide anion), alteration of glutathione levels and its oxidative state, and catalase levels.

2. Detection of changes in the mitochondrial transmembrane potential

Emerging evidence suggests a key role for mitochondria during apoptosis induced by a diverse range of stimuli. Mitochondria not only function as integrators of different pro-apoptotic pathways, but also constitute the target of a number of anti-apoptotic molecules. Discontinuity of the outer mito-

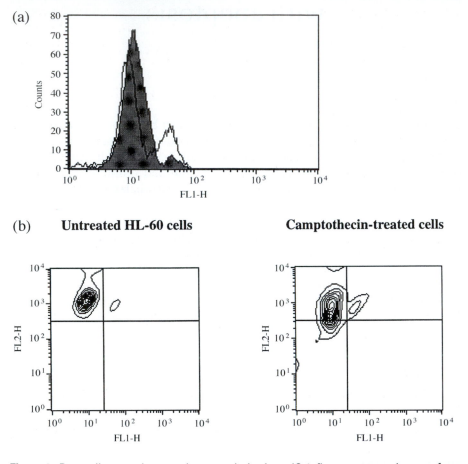

Figure 1. Depending on the membrane polarization, JC-1 fluorescent probe can form monomers (FL-1) or J-aggregates (FL-2). Hence, $\Delta\psi_m$ depolarization can be monitored by measuring the fluorescence in FL-1 and/or FL-2. (a) $\Delta\psi_m$ was measured in untreated DU145 prostate cancer cells (shaded line) and in cells treated with camptothecin (100 ng/ml, 24 h) (solid line), as described in *Protocol 1*. An increase in fluorescence (FL-1) is indicative of membrane depolarization. (b) Treatment of HL60 cells (human promyelocytic leukaemia cell line) with 5.0 µg/ml camptothecin for 6 h causes a significant decrease in the fluorescence measured in FL-2 (seen as a shift to the bottom left panel in the contour plot), which is also indicative of $\Delta\psi_m$ disruption.

chondrial membrane results, in some instances, in redistribution of cytochrome *c* from the intermembrane space to the cytosol, followed by subsequent inner mitochondrial membrane depolarization.

Disruption of $\Delta\psi_m$ is believed to occur through permeability transition (PT), which can be activated in response to a variety of pro-apoptotic signal transduction molecules. The mitochondrial PT pore is a megachannel thought to be directly regulated by the Bcl-2/Bax complex (14). Opening of the PT pore, following PT, allows the release of solutes 1.5 kDa and smaller, and subsequent disruption of $\Delta\psi_m$. As a consequence of intermembrane protein release, caspases and nucleases are released (14). Importantly, inhibitors of PT also inhibit apoptosis in several models of apoptosis, suggesting that disruption of mitochondrial function is of key importance to the apoptotic process (9).

The cell-permeant fluorescent probe 5,5',6,6'-tetrachloro-1,1',3,3'-tetraethyl-benzimidazolylcarbocyanine iodide (JC-1) can be employed to monitor changes in $\Delta\psi_m$ in cells, by flow cytometry. In the presence of a high $\Delta\psi_m$ JC-1 forms what are termed J-aggregates which fluoresce strongly at 590 nm (FL-2). Reduced $\Delta\psi_m$ results in an increased FL-1 signal and/or in a reduced FL-2 signal in JC-1-stained cells (see *Figure 1*). This method allows subpopulations of cells with different mitochondrial properties to be identified (15).

Protocol 1.   Measurement of mitochondrial transmembrane potential ($\Delta\psi_m$) using JC-1

Equipment and reagents

- JC-1 (Molecular Probes), made as a 5 mg/ml stock in DMSO (Sigma). Protect from light and store at –20°C
- FACScan flow cytometer (e.g. Becton Dickinson), with an excitation source of 488 nm

Method

1. Plate the cells in a standard 24-well plate at a density of 5×10^5/ml.

2. Treat the cells with the apoptosis-inducing agents either before, after, or during the incubation period, depending on the time point at which $\Delta\psi_m$ is to be measured.

3. Transfer the sample into a standard FACS tube.

4. Incubate the cells with 5 µg/ml JC-1 for 15 min at 37°C, in the dark.

5. Collect the fluorescence emission through a 530/30 band pass filter (FL-1) and through a 585/42 band pass filter (FL-2), both on a log scale, using a FACScan flow cytometer (see *Figure 1*). For each sample, collect 5000–10 000 events.

3. Detection of intracellular reactive oxygen intermediates (ROI)

Oxidative stress has been suggested as a possible common mediator of apoptosis in response to a number of stimuli (6, 8). Apoptosis can be induced by several agents that are known to cause oxidative stress, or it can be caused directly by the addition of oxidants (7). Accordingly, addition of antioxidants has been shown to block apoptosis induced by a variety of agents and to function at an early stage in the apoptotic process (6). Interestingly, Bcl-2 has been shown to act as an antioxidant (16, 17).

The formation of ROI results from the reduction of oxygen. These ROI include hydrogen peroxide (H_2O_2), superoxide anion ($O_2^{\bullet -}$), and hydroxyl radical ($OH^{\bullet}$) (reviewed in ref. 10). Both superoxide and hydrogen peroxide are relatively unreactive compared with hydroxyl radicals, which can cause damage to most biological molecules. Possible sources of intracellular ROI include the depletion of cellular antioxidants such as GSH, disruption of mitochondrial respiration, and the activation of oxidant-producing enzymes such as NADPH oxidase.

Free radicals can activate tyrosine kinases (18), induce calcium release (19), cause sphingomyelin hydrolysis with consequent ceramide production (20), increase proteolytic degradation of proteins (21), and affect a variety of redox-sensitive targets such as transcription factors (22). Hence, oxidative stress can lead to the activation of other signal transduction pathways.

The fluorescent probes 2',7'-dichlorofluorescin diacetate (DCFH/DA) and dihydroethidium (DHE) may be used for the measurement of intracellular

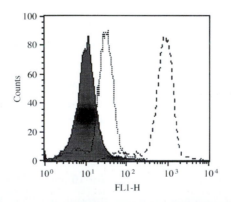

Figure 2. Peroxide levels were assessed in untreated HL60 cells (shaded line) and in cells treated with 5 µg/ml camptothecin for 15 min (dotted line), or cells treated with 1 mM H_2O_2 (broken line) for 30 min, as described in *Protocol 2*. After treatment with camptothecin or with H_2O_2 there is an increase in peroxide production which can be seen as a shift to the right in relative fluorescence (FL-1).

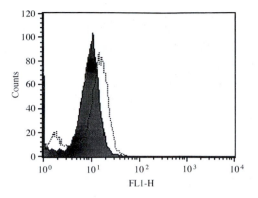

Figure 3. Superoxide levels were assessed in heat-shocked (43°C, 1 h) HL60 cells (dotted line) and in non-heat-shocked cells (shaded line), as described in *Protocol 2*. Heat shock caused an increase in superoxide anion levels in HL60 which can be seen as a shift to the right in relative fluorescence (FL-2).

peroxide and superoxide anion levels, respectively. Both probes can be used in a flow cytometer with a 15 milliwatt, air-cooled, argon ion laser. DCFH/DA is cell permeant and is non-fluorescent until the acetate groups are removed by cellular esterase activity and a peroxide group is subsequently encountered. Hydrolysed, oxidized DFCH/DA fluoresces at 529 nm (FL-1, log scale) (see *Figure 2*) and is unable to leave the cell, thus allowing the measurement of intracellular peroxides by flow cytometry.

DHE is also cell permeant and is oxidized to ethidium by superoxide anion. Once oxidized, ethidium is free to intercalate with DNA in the nucleus, whereupon it emits fluorescence at 605 nm (FL-2; log scale) (see *Figure 3*).

Protocol 2. Measurement of intracellular peroxide levels

Equipment and reagents

- DCFH/DA (Molecular Probes) prepared as a 5 mM stock in DMSO (Sigma). Protect from light and store at –20°C

- FACScan flow cytometer (e.g. Becton Dickinson), with an excitation source of 488 nm

Method

1. Plate the cells in a standard 24-well plate at a density of 5×10^5/ml.

2. Treat the cells with the apoptosis-inducing agents either before, after, or during the incubation period, depending on the time point at which peroxide levels are to be assessed.[a]

3. Transfer the sample into a standard FACS tube.

4. Incubate the cells with 5 μM DCFH/DA, for 1 h at 37°C, in the dark.

Protocol 2. *Continued*

5. Assess the peroxide levels using a FACScan flow cytometer with excitation and emission settings of 488 nm and 530 nm, respectively (FL-1, log scale) (see *Figure 2*). For each sample collect 5000–10 000 events.

[a] As a positive control for peroxide production, cells can be treated with 1 mM H_2O_2 for 30–60 min.

Protocol 3. Assessment of intracellular superoxide anion

Equipment and reagents

- DHE (Molecular Probes) prepared as a 10 mM stock in DMSO (Sigma). Protect from light and store at −20°C

- FACScan flow cytometer (e.g. Becton Dickinson), with an excitation source of 488 nm

Method

1. Plate the cells in a standard 24-well plate at a density of 5×10^5/ml.

2. Treat the cells with the apoptosis-inducing agents either before, after, or during the incubation period, depending on the time point at which the levels of superoxide anion are to be assessed.

3. Transfer the sample into a standard FACS tube.

4. Incubate the cells with 10 μM DHE, for 15 min at 37°C, in the dark.

5. Assess the superoxide levels using a FACScan flow cytometer with excitation and emission settings of 488 nm and 600 nm, respectively (FL-2; log scale) (see *Figure 3*). For each sample collect 5000–10 000 events.

4. Determination of glutathione levels and its oxidative state

Disruption of critically important oxidants can impair the redox status of the cell, leading to oxidative stress. The oxidative changes reported range from decreased levels of reduced glutathione (23, 24), α-tocopherol (23), and protein thiols (23), to downregulation of primary antioxidant defence enzymes such as catalase, manganese superoxide dismutase (MnSOD), Cu/Zn superoxide dismutase (Cu/ZnSOD), and thioredoxin (25).

GSH plays a central role in defending cells against radicals and electrophiles (26). The protective role of GSH consists of four components, namely chemical reaction with intracellular targets, enzymatic reduction of peroxides to prevent their conversion into more reactive species, enzymatic detoxification of electrophiles, and maintenance of the redox state of cellular thiols.

The response of cells to certain anti-tumour drugs and to ionizing radiation can be modulated by alteration of intracellular GSH concentration (27). Indeed, GSH has been implicated in the resistance of certain cell lines to oxygen radicals, various toxins, alkylating agents, and other chemosensitizers (28). In addition, GSH levels have been shown to be greater in many neoplasms compared with normal tissue, thus suggesting that resistance to apoptosis may involve altered levels of GSH (29). Moreover, it has recently been shown (30) that Fas ligation results in a rapid and specific efflux of GSH. Severe depletion of GSH will lower the reducing capacity of a cell, enhancing oxidative stress independent of any increase in the production of ROI.

There are a number of methods (e.g. chemical, enzymatic, chromatographic, fluorimetric) which can be used to measure GSH and/or glutathione disulfide (GSSG). Techniques that assay GSH concentration in a cell extract measure an average concentration, usually expressed as $nmol/10^6$ cells or on a per mg protein basis. Such measurements are not accurate if there is heterogeneity in the population with respect to the GSH content/cell (27). Solid tumours are very heterogeneous in many aspects and, in fact, microenvironmental factors known to be highly heterogeneous in solid tumours are known to influence cellular GSH content (31). In such cases, flow cytometric analysis using monochlorobimane (mBCl) provides more accurate and informative data (described on p. 151).

GSH can be assayed using an enzymatic procedure, which involves its sequential oxidation by 5,5'-dithio-bis(2-nitrobenzoic acid) (DTNB) and reduction by NADPH in the presence of glutathione reductase. The DTNB–glutathione reductase recycling procedure was first described by Tietze (32), and later modified by Griffith (33). GSH is oxidized by DTNB to stoichiometrically give GSSG and 5-thio-2-nitrobenzoic acid (TNB) (see *eqn 1*). GSSG is then reduced to GSH by glutathione reductase and NADPH (see *eqn 2*). The rate of TNB formation can be followed spectrophotometrically (at 412 nm, or 405 nm) and it is proportional to the sum of GSH and GSSG present. The assay can also be monitored at 340 nm (NADPH).

$$2GSH + DTNB \rightarrow GSSG + TNB \qquad (1)$$
$$GSSG + NADPH \rightarrow 2GSH + NADP^+ \qquad (2)$$

One of the advantages of the enzymatic assay is its ability to measure extremely low GSH levels provided that there is enough starting material. It is also a specific and reliable procedure. It is critical, however, to set up appropriate standards, as the method depends on an accurate standard curve.

Because GSSG is normally present at very low concentrations compared with GSH, the determination of GSSG is often difficult. The above assay can be made specific for GSSG by masking any GSH present with *N*-ethylmaleimide (NEM) (32). Although this procedure is effective, residual NEM must be removed as it is a potential inhibitor of glutathione reductase. This process is not only time consuming but also a potential source of error (33). The method described by Griffith (33) uses 2-vinylpyridine (2-VP), a reagent which does not inhibit glutathione reductase and thus need not to be removed. The GSH present in solutions of pH >5.5 is readily derivatized by adding 2 μl neat 2-VP/100 μl solution and mixing the solution for 1 min. Depending on the final pH, GSH (up to at least 15 mM) will be fully derivatized after 20–60 min at 25°C. Acid solutions must be at least partially neutralized (e.g. with triethanolamine, which is very soluble and, due to its pK_a, is unlikely to increase the pH to the point where auto-oxidation is rapid) before the reaction of GSH and 2-VP. In some circumstances the 2-VP added is sufficient to raise the pH.

Protocol 4. DTNB–glutathione reductase recycling assay for GSH and GSSG

Reagents[a]

- PBS
- 5-sulfosalicylic acid (SSA) (Sigma)
- stock buffer: 125 mM sodium phosphate, 6.3 mM Na_4EDTA, pH 7.5
- DTNB (Sigma), is prepared in stock buffer as a 6 mM stock solution

- NADPH (Sigma), prepared fresh as a 0.3 mM stock solution, in stock buffer
- glutathione reductase (Sigma). The yeast enzyme is diluted to 50 U/ml in stock buffer
- GSH standards (from 100 mM stock solution), are diluted daily in 3% (w/v) SSA

Method

1. Harvest the cells (5×10^6) from culture by centrifugation (200 *g*, 5 min).

2. Wash the cells by resuspending in 4 ml of PBS and centrifuging as described in step 1.

3. Lyse the cells in 1.2 ml water.

4. Mix 900 μl lysate with 100 μl 30% (w/v) SSA, and incubate for 15 min on ice.

5. Centrifuge the sample at 12000 *g* for 2 min.

6. In a 1 cm light path cuvette, mix 700 μl of NADPH solution (0.21 mM), 100 μl of DTNB (0.6 mM) solution, and 100 μl GSH sample[b] (or water) to give a final volume of 1 ml.

7. Equilibrate the cuvette to 30°C (e.g. water bath or incubator), for 15 min.

8. To initiate the assay, add 10 μl glutathione reductase (0.5 U/ml) and follow the formation of TNB by monitoring the absorbance at 412 nm, until it exceeds 2.0.[c]

---

9. Determine the amount of GSH from a standard curve in which the GSH equivalents are plotted against the rate of change in A_{412}. Express the level of GSH as GSH equivalents (e.g. nmol/10^6 cells).

[a] All solutions are stable for two weeks at 0°C, although the NADPH solution forms a precipitate which must be dissolved by warming to room temperature.
[b] As a blank, use a sample lacking GSH.
[c] The linear portion of the curve is usually between 1 and 2 absorbance units.

Protocol 5.   DTNB–glutathione reductase recycling assay for GSSG

Reagents[a]

- PBS
- SSA, prepared as a 30% (w/v) solution
- 2-VP (Sigma), used undiluted and stored at −20°C. Replace 2-VP if it becomes viscous or brownish
- triethanolamine (Sigma), used undiluted
- stock buffer: 125 mM sodium phosphate, 6.3 mM Na₄EDTA, pH 7.5

- DTNB (Sigma), is prepared in stock buffer as a 6 mM stock solution
- NADPH (Sigma), prepared fresh as a 0.3 mM stock solution, in stock buffer
- glutathione reductase (Sigma). The yeast enzyme is diluted to 50 U/ml in stock buffer
- GSSG standards (e.g. 50 mM stock) (Sigma), are diluted daily in 3% (w/v) SSA

Method

1. Harvest the cells (5×10^6) from culture by centrifugation (200 *g*, 5 min).

2. Wash the cells by resuspending in 4 ml of PBS and centrifuging as described in step 1.

3. Lyse the cells in 1.2 ml water.

4. Mix 900 μl lysate with 100 μl 30% (w/v) SSA, and incubate for 15 min on ice.

5. Centrifuge the sample at 12 000 *g* for 2 min.

6. Add 2 μl of 2-VP to a 100 μl SSA supernatant aliquot with gentle agitation.[b,c]

7. Add 6 μl (45 μmol) of neat triethanolamine to the side of the test tube, above the liquid level, and agitate the tube vigorously.[d]

8. Check the pH and make sure it is between 6 and 7.[e]

9. Allow the samples to stand at room temperature for 60 min.

10. In a 1 cm light path cuvette, mix 700 μl of NADPH solution (0.21 mM), 100 μl of DTNB (0.6 mM) solution, GSSG sample (108 μl), and 92 μl water to give a final volume of 1 ml.

11. Equilibrate the cuvette to 30°C (e.g. water bath or incubator), for 15 min.

Protocol 5. *Continued*

12. To initiate the assay, add 10 μl glutathione reductase (0.5 U/ml) and follow the formation of TNB by monitoring the absorbance at 412 nm, until it exceeds 2.0.

13. Determine the amount of GSSG from a standard curve.

a All solutions are stable for two weeks at 0°C, although the NADPH solution forms a precipitate which must be dissolved by warming to room temperature.
b The derivatization procedure should be carried out in a fume hood, as 2-VP has a low vapour pressure and frequent exposure to it might be irritating.
c Run a blank containing 2-VP, but no GSSG.
d Since triethanolamine gives a small interference with the assay, the standards should contain an amount of amine equivalent to the samples.
e If the pH inadvertently exceeds 7, redo derivatization on a new sample using less triethanolamine.

Several fluorescent reagents have been developed for determining cellular levels of GSH. However, no probe is without drawbacks in quantitative studies of live cells. The high (up to 10 mM) but variable levels of intracellular GSH make kinetic measurements under saturating substrate conditions difficult (34, 35). The fluorescent reagents designed to measure GSH may react with other intracellular thiols, including proteins in GSH-depleted cells (36). Monochlorobimane (mBCl) is a reagent that reacts non-enzymatically with GSH. It is cell-permeant and non-fluorescent until conjugated. The fluorescent adduct formed between mBCl and GSH can be detected by cytofluoro-

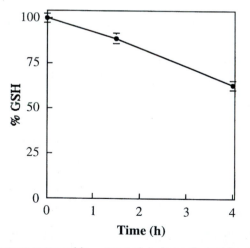

Time (h)

Figure 4. GSH levels were mesured in untreated Jurkat cells and in cells treated with 200 ng/ml anti-Fas IgM, for various periods of time. Following anti-Fas treatment there is a decrease in cellular GSH content, which was measured as described in *Protocol 6*. The data (from triplicate samples) represent mean %GSH±SE of each sample in relation to the control/untreated sample (100% GSH) and results shown are representative of three independent experiments.

metry, or by flow cytometry using 50 milliwatt UV excitation and emission integrated above 425 nm. While mBCl will react non-specifically with many different thiols, preferential derivatization of GSH can be achieved by using a low concentration of mBCl (short time incubation), as the reaction with GSH is catalysed by GST and the non-enzymatic reaction is very slow (27).

Although the cytofluorimetric assay is not quantitative, it will allow the comparison of relative GSH content in different samples (e.g. GSH content in cancer cells before and after treatment with a particular chemotherapeutic drug) (see *Figure 4*). In addition, the assay is reliable, easy, and rapid to perform.

Protocol 6. GSH determination with monochlorobimane (mBCl) by cytofluorometry

Equipment and reagents
- mBCl (Molecular Probes), prepared as 10 mM stock in ethanol. Protect from light and store at –20 °C
- PBS
- cytofluorimeter (e.g. SLM-AMINCO)

Method

1. Harvest 1×10^6 and centrifuge at 200 g for 5 min at room temperature.

2. Wash the cells by resuspending in 4 ml of PBS and centrifuging as described in step 1.

3. Resuspend the pelleted cells in 2 ml of PBS.

4. Incubate the cells with 50 μM of mBCl for 15 min at room temperature, in the dark.[a]

5. Read the fluorescence of the sample with excitation set at 395 nm and fluorescence emission set at 482 nm (see *Figure 4*).

[a] As a blank use 2 ml of PBS incubated with 50 μM mBCl for 15 min at room temperature.

The flow cytometric method was first described by Rice *et al.* (31). It was shown that the GSH–bismane adduct, quantitated by HPLC, was formed with similar kinetics to the appearance of cellular fluorescence, as quantitated by flow cytometry. In addition, analysis of GSH by flow cytometry showed a strong correlation with parallel GSH analysis by an enzymatic method over a wide range of values for intracellular GSH. Hence, derivatization of cellular GSH with mBCl allows measurement of single-cell GSH content by flow cytometry. The kinetics of the reaction should be determined for each cell type to be sure that the reaction goes to completion, and the stoichiometry of the reaction should also be taken into account (27). For example, if the GSH content of 10^6 cells is 10 nmol, then at least 10 μM mBCl would be required to derivatize cellular GSH at a density of 10^6 cells/ml, and the concentration of mBCl would change as derivatization progresses. The assay does not require large numbers of cells but may not be applicable to all cell types (e.g. cells with low levels of glutathione *S*-transferase (GST). Although the assay may not discriminate very low GSH levels, it will reveal subpopulations and the fractions of cells contained in each, as well as their relative GSH contents.

Protocol 7. Flow cytometric analysis of GSH using monochlorobimane (mBCl)

Equipment and reagents

- mBCl (Molecular Probes), prepared as 10 mM stock in ethanol. Protect from light and store at –20°C
- flow cytometer equipped with a 50 milliwatt UV excitation and emission integrated above 425 nm

Method

1. Harvest 1×10^6/ml cells into a standard FACS test tube.

2. Stain the cells with 40 µM mBCl at room temperature for 5 min and immediately transfer the cells to ice.

3. Read samples using a flow cytometer with excitation emission integrated above 425 nm. For each sample collect 5000–10 000 events.

DL-buthionine-*S,R*-sulfoximine (BSO) is a specific inhibitor of GSH synthesis, which has been shown to dramatically reduce GSH levels (28) (see *Figure 5*). The disruption of antioxidant mechanisms in cells can impair the

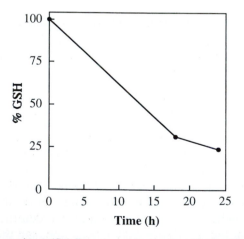

Figure 5. GSH levels can be artificially reduced by treating HL60 cells with 100 µg/ml BSO (see *Protocol 8*) for various periods of time. GSH levels were assessed as described in *Protocol 6*. The data (from triplicate samples) represent mean %GSH±SE and the results shown are representative of three independent experiments.

redox status of the cell, leading to oxidative stress and subsequent apoptosis. For instance, hormone-mediated apoptosis in prostate tissue is preceded by an increase in GST expression (37). McGowan *et al.* (38) have shown that artificial depletion of GSH significantly increased the levels of peroxide in leukaemic cells and caused subsequent apoptosis.

Protocol 8. Reduction of glutathione levels in cell lines with DL-buthionine-*S,R*-sulfoximine (BSO), a specific inhibitor of GSH synthesis

Reagents
- BSO (Sigma) freshly prepared as a 1 mg/ml stock in cell culture media (e.g. RPMI 1640)

Method

1. Plate the cells in a standard 24-well plate at a density of 5×10^5/ml.

2. Incubate the cells, at 37°C, with varying concentrations of BSO (e.g. 100 μg/ml), for various time intervals (e.g. 18–24 h), in order to determine the optimal concentration of BSO.

3. Assess the levels of GSH as previously described (see *Protocols 4, 6* or *7*, and *Figure 5*).

---

5. Measurement of catalase in cells undergoing apoptosis

Catalase has a dual function: (a) dismutation of hydrogen peroxide (H_2O_2) to oxygen and water (catalytic activity) (see *eqn 3*) and (b) oxidation of hydrogen donors such as methanol, ethanol, formic acid, and phenols, with the decomposition of 1 mol of peroxide (peroxidic activity) (see *eqn 4*) (39).

$$2H_2O_2 \rightarrow 2H_2O + O_2 \tag{3}$$
$$ROOH + AH_2 \rightarrow H_2O + ROH + A \tag{4}$$

Because the kinetics of catalase do not obey a normal pattern, measurements of the enzyme activity at substrate saturation, or determination of the K_s is virtually impossible (40). It is not possible to saturate the enzyme with substrate within the physiological range; however, at H_2O_2 concentrations above 0.1 M there is a rapid inactivation of catalase. Hence, to avoid a rapid decrease in the initial rate of the reaction, the spectrophotometric assay described below [which is fundamentally that of Aebi (40)] must be carried out at low H_2O_2 concentration. The decomposition of H_2O_2 can be followed directly by the decrease in the absorbance measured at 240 nm in a spectrophotometer. It follows initially (0–30 sec) a first-order reaction, with H_2O_2

concentration between 0.01–0.05 M. The decrease in absorbance per unit time is a measure of catalase activity. Owing to the abnormal kinetics it is not possible to define international catalase units. Thus, one unit of catalase is defined as the amount of enzyme that catalyses the decomposition of 1 μmol H_2O_2/min. The calculation of activity (U/ml) can be determined using *eqn 5:*

$$\text{activity} = (\Delta A_{240}/t)/\varepsilon \times \text{assay volume} \times 1000/\text{sample volume} \qquad (5)$$

where $\Delta A_{240}/t$ is the decrease in the absorbance at 240 nm measured spectrophotometrically over a period of time and ε is the extinction coefficient of catalase (40 mM^{-1}cm^{-1}). The specific activity of the enzyme can be obtained by dividing the activity by the protein concentration in the sample (U/mg protein).

Protocol 9.   Measurement of catalase in cells

Reagents

- Assay buffer: 50 mM phosphate buffer pH 7.0
- H_2O_2 (BDH) is diluted prior to the assay with assay buffer to 30 mM

Method

1. Harvest by centrifugation (200 *g*, 5 min) 6–10 × 10^6 cells.
2. Resuspend the cells in ~600 μl assay buffer and freeze the cells at −20°C, until the experiment is carried out.
3. Thaw the cells at room temperature and refreeze at −70 °C.
4. Freeze–thaw the cells twice more, as described in step 3.[a]
5. Centrifuge the cells at 200 *g* for 5 min.
6. In a 1 cm light path cuvette, mix 1 ml assay buffer and 500 μl supernatant.
7. To initiate the reaction add 1 ml H_2O_2 to the cuvette and monitor the absorbance at 240 nm over 30 sec.[b]

[a] Once diluted, catalase is not a very stable enzyme, therefore the assay should be carried out within 10 min of the final freeze–thaw procedure.
[b] The blank should contain 2 ml assay buffer and 500 μl supernatant.

Acknowledgements

This work has been supported by the Foundation for Science and Technology (Fundação para a Ciência e a Tecnologia), Lisbon, Portugal and the Health Research Board of Ireland. The authors would like to thank Emma Creagh and George Royle for many helpful discussions.

References

1. Kerr, J.F.R., Wyllie, A.H. and Currie, A.R. (1972). *Br. J. Cancer*, **26**, 239.
2. Wyllie, A.H., Kerr, J.F.R. and Currie, A.R. (1980). *Int. Rev. Cytol.*, **68**, 251.
3. Cohen, J.J., Duke, R.C., Fadok, V.A. and Sellins, K.S. (1992). *Annu. Rev. Immunol.*, **10**, 267.
4. Bortner, C.D., Oldenburg, N.B.E. and Cidlowski, J.A. (1995). *Trends Cell Biol.*, **5**, 21.
5. Martin, S.J. and Green, D.R. (1995). *Cell*, **82**, 349.
6. Buttke, T.M. and Sandstrom, P.A. (1994). *Immunol. Today*, **15**, 7.
7. Lennon, S.V., Martin, S.J. and Cotter, T.G. (1991). *Cell Prolif.*, **24**, 203.
8. McGowan, A.J., Ruiz-Ruiz, M.C., Gorman, A.M., Lopez-Rivas, A. and Cotter, T.G. (1996). *FEBS Lett.*, **392**, 299.
9. Kroemer, G. (1997). *Cell Death Differ.*, **4**, 443.
10. Gorman, A.M., McGowan, A.J., O'Neill, C. and Cotter, T.G. (1996). *J. Neurol. Sci.*, **139**, 45.
11. Staal, F.J.T., Ela, S.W., Roederer, M., Anderson, M.T. and Herzenberg, L.A. (1992). *The Lancet*, **2**, 1294.
12. Briehl, M.M. and Baker, A.F. (1996). *Cell Death Differ.*, **3**, 63.
13. Cerrutti, P.A. (1985). *Science*, **227**, 375.
14. Kroemer, G. (1998). *Cell Death Differ.*, **5**, 547.
15. Salvioli, S., Ardizzoni, A., Franceschi, C. and Cossarizza, A. (1997). *FEBS Lett.*, **411**, 77.
16. Hockenbery, D.M., Oltvai, Z.N., Yin, X.M., Milliman, C.L. and Korsmeyer, S.J. (1993). *Cell*, **75**, 241.
17. Zhong, L.T., Sarafian, T., Kane, D.J., Charles, A.C., Mah, S.P., Edwards, R.H. and Bredesen, D.E. (1993). *Proc. Natl. Acad. Sci.*, **90**, 4533.
18. Bauskin, A.R., Alkalay, I. and Ben-Neriah, Y. (1991). *Cell*, **66**, 685.
19. Richter, C. (1993). *FEBS Lett.*, **325**, 104.
20. Jayadev, S., Linnardic, C.M. and Hannun, Y.A. (1994). *J. Biol. Chem.*, **269**, 5757.
21. Grune, T., Reinheckel, T., Joshi, M. and Davies, K.J.A. (1995). *J. Biol. Chem.*, **270**, 2344.
22. Schreck, R., Rieber, P. and Baeuerle, P.A. (1991). *EMBO J.*, **10**, 2247.
23. Slater, A.F.G., Nobel, C.S.I., Maellaro, E., Bustamante, J., Kimland, M. and Orrenius, S. (1995). *Biochem. J.*, **306**, 771.
24. Beaver, J.P. and Waring, P.A. (1995). *Eur. J. Cell Biol.*, 68, 47.
25. Briehl, M.M., Cotgreave, I.A. and Powis, G. (1995). *Cell Death Differ.*, **2**, 41.
26. Bump, E.A., Yu, N.Y., Taylor, Y.C., Brown, J.M., Travis, E.L. and Boyd, M.R. (1983). In *Radioprotectors and anticarcinogens* (ed. M.G. Simic and O.F. Nygaard), p. 297. Academic Press, New York.
27. Shrieve, D.C., Bump, E.A. and Rice, G.C. (1988). *J. Biol. Chem.*, **263**, 4107.
28. Green, J.A., Vistica, D.T., Young, R.C., Hamilton, T.C., Rogan, A.M. and Ozols, R.F. (1984). *Cancer Res.*, **44**, 5427.
29. Hamilton, T.C., Winker, M.A., Louie, K.G., Batist, G., Behrens, B.C., Tsuruo, T., Grotzinger, K.R., McKoy, W.M., Young, R.C. and Ozols, R.F. (1986). *Int. J. Radiat. Oncol. Biol. Phys.*, **12**, 1175.
30. Van den Dobbelsteen, D.J., Nobel, C.S.I., Schlegel, J., Cotgreave, I.A., Orrenius, S. and Slater, A.F.G. (1996). *J. Biol. Chem.*, **271**, 15420.

31. Rice, G.C., Bump, B.A., Shrieve, D.C., Lee, W. and Kovacs, M. (1986). *Cancer Res.*, **46**, 6105.
32. Tietze, F. (1969). *Anal. Biochem.*, **27**, 502.
33. Griffith, O.W. (1980). *Anal. Biochem.*, **106**, 207.
34. Van der Ven, A., Mier, P., Peters, W.H., Dolstra, H., Van Erp, P.E., Koopmans, P.P. and Van der Meer, J.W. (1994). *Anal. Biochem.*, **217**, 41.
35. Hedley, D.W. and Chow, S. (1994). *Cytometry*, **15**, 349.
36. Nair, S., Singgh, S.V. and Krishan, A. (1991). *Cytometry*, **12**, 366.
37. Jung, K., Seidel, B., Rudolph, B., Lein, M., Cronaeur, M.V., Henke, W., Hampel, G., Schnorr, D. and Loening, S.A. (1997). *Free Radic. Biol. Med.*, **23**, 127.
38. McGowan, A.J., Bowie, A.G., O'Neill, L.A.J. and Cotter, T.G. (1998). *Exp. Cell Res.*, **238**, 248.
39. Chance, B. (1947). *Acta Chem. Scand.*, **1**, 236.
40. Aebi, H. (1984). In *Methods in enzymology* (ed. L. Packer), Vol. 105, p. 121. Academic Press, London.

9

Methods of measuring Bcl-2 family proteins and their functions

JOHN C. REED, ZHIHUA XIE, SHINICHI KITADA,
JUAN M. ZAPATA, QUNLI XU, SHARON SCHENDEL,
MARYLA KRAJEWSKA, and STANISLAW KRAJEWSKI

1. Introduction

Bcl-2 family proteins play critical roles in the regulation of programmed cell death and apoptosis (reviewed in refs 1–4). Changes in the levels or bioactivities of these proteins are associated with a variety of physiological processes where cell death occurs, including fetal development, haemato-poietic and immune cell differentiation, oogenesis, mammary gland involution, and normal cell turnover in the epidermis, gut, and several other tissues. Moreover, pathological alterations in the expression of Bcl-2 family proteins have been documented in cancer, autoimmunity, immunodeficiency, heart failure, stroke, and other diseases (reviewed in refs 5–9). Consequently, methods for assessing the relative levels and bioactivities of Bcl-2 family proteins can be of interest to scientists in a board range of disciplines.

Here, we describe some of the methods routinely employed in our laboratory for this purpose. Many of the techniques described are specifically intended for analysis of the Bcl-2 (anti-apoptotic) or Bax (pro-apoptotic) proteins or mRNAs, but can be modified readily for applications to other members of the Bcl-2 family.

2. Immunoblot analysis

The analysis of Bcl-2 family proteins by immunoblotting involves: (a) lysis of cells or processing of tissues for extraction and recovery of intact proteins; (b) SDS–PAGE separation of proteins and transfer to either nitrocellulose or nylon membranes; and (c) detection of antigens using antibody-based methods. In addition, the reprobing of blots with additional antibodies is often an important consideration. Here, we describe methods which have provided successful for extraction of Bcl-2 family proteins from cultured cells and tissues, and for immunodetection of these proteins once immobilized on membranes. A procedure developed in our laboratory which permits multiple reprobings of blots is also described (10).

2.1 Cell lysis and tissue processing procedures

2.1.1 Cultured cells

Protocol 1.

Method

1. (a) Suspension cells are collected by centrifugation from culture medium. (b) For adherent cells, remove the culture medium completely from the culture dish. Add 5 ml of PBS and shake gently to wash cells. Then remove PBS completely. Add 5 ml of ice-cold PBS containing PMSF and scrape off with the aid of a rubber policeman (do not trypsinize adherent cells). Transfer cells to a 15 ml centrifuge tube and centrifuge at 500 g for 5 min.
2. Remove the supernatant carefully from cell pellets and add 1.5 ml of ice-cold PBS/PMSF. Pipette cells gently to resuspend and transfer the cells to a 1.5 ml microcentrifuge tube.
3. Centrifuge at 500 g for 0.5–1 min with a swinging bucket rotor at 4°C.
4. Remove the supernatant carefully and add either TEN-T solution (described on p. 160) or RIPA buffer with protease inhibitors. Pipette to resuspend the cell pellet. The volume of lysis solution depends on the cell numbers harvested. We usually add 200 µl of TEN-T solution to 1 × 10^7 cultured cells.
5. Keep on ice for 15 min. Then centrifuge at 10 000 g for 20 min. at 4°C (in the cold).
6. Recover supernatant and store at –80°C.

2.1.2 Tissues

Tissue specimens are either frozen in liquid nitrogen (N_2) and tissue powder is prepared by crushing in a mortar and pestle in N_2 followed by solubilization with a lysis buffer, or fresh tissue pieces are mechanically homogenized and sonicated in lysis buffers.

Fresh tissues

Protocol 2.

Method

1. Cut the tissue sample with scalpel blades into small pieces and transfer them to 1.5 ml Eppendorf tube.

---

2. Immediately add RIPA[a] lysis buffer to the tissue and homogenize using a motor-driven pestle (e.g. KONTES Scientific Glassware/Instruments, Vineland, NJ[b]). The volume of lysis buffer depends on the volume of sample harvested. We usually add lysis buffer at a 5:1 ratio of buffer to tissue volume (5 vol. RIPA buffer:1 vol. tissue).

3. Incubate on ice for 15 min.

4. Centrifuge 800–1600 *g* for 20 min at 4°C to pellet debris.

5. Recover the supernatant and store at –80°C.

[a] An alternative to RIPA solutions is to use a modified Laemmli solution. Resuspend tissues in boiling 1 × Laemmli buffer lacking bromophenol blue dye and 2-mercaptoethanol at a 5:1 ratio of buffer to tissue volume. Homogenize and centrifuge at 800–1600 *g* for 20 min at room temperature (SDS will precipitate at 4°C). Store the supernatant at –20°C.
[b] Alternatively, disrupt tissue using a probe sonicator. A 25 μm probe and ~4 cycles of 0.5 minutes each typically works well. (Avoid heating lysates).

Frozen tissues

Frozen tissue specimens may either be derived from tissue samples that were placed directly in liquid nitrogen (and stored in the same or at –80°C) or derived from histological specimens which were frozen in OTC compound (outside tissue compound, Tissue Tek-II, USA, VWR) for cryosectioning. The latter typically involves submerging in cold isopentane which is sitting in a dry-ice–acetone bath. The frozen material is then placed into a pre-cooled plastic mould and OTC compound is poured over the tissue, thus encasing it in a 'plastic-like' covering. The samples are then stored at –80°C and can be shipped on dry-ice.

If samples are embedded in OTC, this covering must first be removed. Place the tissue specimen in a petri dish that has been pre-cooled on dry-ice. Working quickly, use scalpel and forceps to remove the OTC coating.

Protocol 3.

Method

1. Immediately submerge the tissue in liquid nitrogen in a pre-cooled mortar and pestle.

2. Grind the tissue to a fine powder and transfer with the remaining liquid N_2 into a polypropylene centrifuge tube. (**Caution.** Do not close the tube completely or it will blow up! Close the cap partially to prevent tissue from bubbling out and hold in a partially closed position with a gloved finger until 'boiling' of the liquid N_2 subsides.)

3. When the liquid nitrogen has nearly all evaporated, add 1–2 volumes (at least!) of RIPA buffer with protease inhibitors.

4. This will initially freeze. Let it thaw on ice.

5. Vortex vigorously.

Protocol 3. *Continued*

6. Sonicate (e.g. Heat-Systems Ultrasonicator using the standard 1.5 mm diameter tip and 1–3 sec pulses and power settings of 2–4). If the sample heats beyond ~ room temperature, put on ice and allow to cool. Sonicate until the solution is no longer viscous and has become clear, without residual chucks of tissue.

7. Centrifuge in a microcentrifuge (16 000 *g*) for 20 min and transfer the supernatant to a fresh microcentrifuge (1.5 ml) tube.

8. Store the supernatant at –80 °C or keep on ice and proceed directly to protein quantification and subsequent SDS–PAGE/immunoblot analysis.

2.1.3 Protein quantification

To quantify the total protein concentration in lysates, we use the bicinchoninic acid (BCA) protein assay reagent (PIERCE) for Triton X-100-solubilized RIPA lysates (11) and for Laemmli lysates, provided β-mercaptoethanol has been withheld until ready to boil, and load samples in gels.

Use the Coomassie protein assay reagent (PIERCE) for Laemmli lysates because SDS interferes with the BCA method.

2.1.4 Lysis solutions

TEN-T solution. This solution is generally reliable for extraction of Bcl-2 family protein from most cultured cells. A modified version is useful for co-immunoprecipitation experiments in which 0.2–1 % (v/v) NP-40 is substituted for Triton X-100. Phosphatase inhibitors are optional, depending on whether experiments entail assessment of phosphorylation of Bcl-2 family proteins.

150 mM NaCl
1% Triton X-100
10 mM Tris (pH 7.4)
5 mM EDTA (pH 8.0)

Make from autoclaved solutions. Store at room temperature but add protease inhibitors fresh each time to the aliquot of the solution needed for your experiments.

RIPA solution. This lysis solution reliably extracts Bcl-2 family proteins for cultured cells and many tissues, including frozen tissues. It can be stored at room temperature and protease inhibitors are added immediately prior to use. Phosphatase inhibitors are optional, depending on whether the experiments entail assessment of phosphorylation of Bcl-2 family proteins.

10 mM Tris (pH 7.4)
150 mM NaCl
1% Triton X-100

1% deoxycholate

0.1% SDS

5 mM EDTA

Laemmli solution. Ensures vigorous extraction of proteins from tissue specimens.

6.25 ml of 0.5 M Tris (pH 6.6)

11.25 ml of 10% SDS

5 ml of glycerol

Bring volume to 45 ml

Store at room temperature

When used as a loading solution for SDS–PAGE, the solution is typically prepared at 2 × concentration and mixed with an equal volume of sample lysate. In this case, BPB (bromophenol blue; 5 mg per 45 ml) is added and 10% 2-mercaptoethanol is included fresh each time to the portion needed. *Protease inhibitors.* Table 1 lists the protease inhibitors.

Phosphatase inhibitors (final concentrations).

10 mM sodium β-glycerophosphate (pH 7.4)

1 mM Na_3VO_4 (sodium orthovanadate)

5 mM NaF

2 mM $Na_4P_2O_7$ (pyrophosphate)

50 mM 4-nitrophenyl phosphate

1 μM microcystin-LR

2.2 Immunoblotting procedure

The methods for SDS–PAGE separation and transfer of Bcl-2 family proteins to membranes are not unusual, and thus detailed protocols are not provided here. We have, however, devised an immunoblotting method which permits sequential detection of multiple antigens on a single protein blot without

---

Table 1. Protease inhibitors

Protease inhibitor	Stock solution	Solvent	Dilution
PMSF (phenylmethyl-sulfonyl fluoride)	100 mM	Methanol	1:100
Aprotinin (Sigma, cat. no. A-6279)	28 TIU /ml	PBS	1:100
Leupeptin	5 mg/ml	PBS	1:100
Benzamidine	100 mM	PBS	1:100
Pepstatin	0.7 mg/ml	Methanol	1:1000

All of these are stored at –20°C in 1 ml aliquots, except aprotinin solution which is stored at 4°C.

---

stripping off prior antibodies (10). The procedure is described here. It has worked well in our hands for detection of multiple members of the Bcl-2 family of proteins, using antibodies generated in our laboratory or from commercial sources.

The multiple antigen detection (MAD) procedure utilizes horseradish peroxidase (HRPase)-based detection with a chemiluminescent substrate. After detection by enhanced chemiluminescence (ECL) with exposure to X-ray film, the antigen–antibody complexes on the blot are treated with a chromogenic substrate [either 3,3'-diaminobenzidine (DAB) or the "SG" substrate (Vector Labs, Inc.)] which renders the antigen–antibody–HRPase complexes unreactive in subsequent reprobings of the same membrane with additional antibodies using the same detection method. Because no stripping is involved, immobilized proteins are not lost from the membrane, thus allowing for multiple sequential reprobings of the same membrane with different primary antibodies (>12) and retention of strong signal intensities for sequential antibody probings.

2.2.1 Multiple antigen detection (MAD) procedure

SDS–PAGE is typically performed using 12% gels, loading either 50–100 or 10–25 μg of total protein per lane, depending on whether one is using a standard size gel (e.g. 10 cm × 14 cm) or a minigel (4 cm × 6 cm). Cell lysates should be mixed with an equal volume of 2 × Laemmli buffer containing 10% 2-mercaptoethanol (Sigma Inc.) and boiled for 5 min prior to loading in gels. Proteins are transferred from gels to either nitrocellulose membranes (0.45 μm; Biorad) or PVDV nylon membranes (Amersham Life Sci) using Tris–glycine buffer [25 mM Tris (pH 8.3), 192 mM glycine, 20% methanol, 0.01 % SDS]. Typically, we transfer at 280 mA for 2 h when working with minigels and at 350 mA for 3.5 h when preparing standard-size gels.

Protocol 4.

Method

11. Wash the blot for 5 min in modified PBS (mPBS) solution using at least 0.5 ml per cm² blot.

All incubation steps described here should be performed with gentle agitation on a rotating or rocking platform shaker at room temperature (unless an overnight incubation is involved, in which case 4°C should be used).

12. Transfer the nitrocellulose membranes to ~0.25–0.5 ml/cm² pre-blocking solution (TN-TBM) with optional 1% normal goat serum and incubate for 1 h at room temperature.

13. Decant off the pre-blocking solution and save (can be stored at –20°C and reused ~10 times).

---

14. Incubate the blot overnight at 4°C in ~0.25 ml/cm² TN-TBM (see *, p. 165, 2.2.3) containing 0.2% normal goat serum (or serum from whatever species the secondary antibody was raised in) and primary antibody at an empirically determined appropriate dilution [most high-titre rabbit polyclonals can be used at ~1:1000~1:5000 (v/v), whereas monoclonals are employed at ~0.1–1.0 μg/ml].

For antisera raised using ovalbumin as the carrier protein, also add 50 μg/ml of ovalbumin (Sigma, Inc.) to the TN-TBM.

15. Decant off the primary antibody solution and save (with high-titre polyclonal antisera, this solution can be stored at –20°C and reused typically ~5–10 times).

16. Rinse the blot briefly in mPBS and then wash three times for 30 min each with ≥0.5 ml/cm² of mPBS.

17. Incubate the blot for 2 h to overnight in ~0.25–0.5 ml/cm² of TN-TBM containing 0.2% goat serum (or other appropriate non-immune serum from the same species as the source of the secondary antibody), and secondary antibody.

The secondary antibody detection method can involve use of either HRPase-conjugated anti-Ig as a single-step reagent, or biotinylated anti-Ig with subsequent incubation with HRPase–avidin–biotin complex (ABC) reagent. For HRPase-conjugated anti-Ig, we use affinity-purified reagents at 1:3000 (v/v) (~0.65 μg/ml) for goat anti-rabbit Ig and goat anti-mouse Ig and at 1:3000 (v/v) (~0.2 μg/ml) for sheep anti-mouse Ig. For avidin–biotin-based detection, we typically employ biotinylated goat anti-rabbit IgG at 1–1.5 μg/ml, anti-mouse IgG at 0.6–1.0 μg/ml, or anti-mouse IgM at 0.6–1.0 μg/ml.

18. Decant off the secondary antibody solution and save at –20°C (can be reused ~2–5 times), rinse the blots once with PBS, and wash three times for 30 min each in =0.5 ml/cm² of mPBS.

19. If using the avidin–biotin method, prepare HRPase–ABC reagent by dispensing two drops of Vector Labs solution A into 5 ml 0.1 M Tris (pH 8.2), mixing, and then adding two drops of the manufacturer's solution B, and allowing the complex to form for 1 h. Incubate the washed blot in 0.25–0.5 ml/cm² of HRPase–ABC solution for 45 min to 1 h. Save the ABC solution at –20°C (can be reused ~10 times).

10. Rinse in PBS and wash the blot three times for 10–30 minutes each in mPBS.

11. Antibody detection is then accomplished using a chemiluminescent substrate reagent (Amersham, Inc.) with exposure to X-ray film (Eastman-Kodak, Inc., XOMAT-AR) according to the manufacturer's instructions. Eliminate excess fluid from the membrane prior to ECL, then place the blot into a plastic bag and add just enough ECL reagent to ensure complete coverage of the blot surface. Immediately expose

Protocol 4. *Continued*

to X-ray film in the dark. Exposure times vary widely, from a few seconds to 5 minutes. We typically develop after 15–30 secs initially, and then adjust the exposure time based on this result.

12. Rinse the blot in PBS for 3 min, and add 0.25 ml/cm² 'activated' SG chromogenic substrate (Vector Labs, Inc.). For this purpose, the SG substrate is prepared by dispensing three drops of Vector SG substrate into 15 ml of PBS, mixing well, and then adding three drops of H_2O_2 (0.5%) and mixing. DAB is also suitable but tends to cause more damage to blots and thus reduces the number of reprobings possible.

It is important to designate containers in the laboratory for SG substrate use and not to use these for ECL. The SG substrate is very difficult to wash off plastic and will compete with ECL reactions. (Bleach, however, can remove most SG substrate from plastic containers.)

13. Incubate for 15–30 min.

14. Discard the substrate solution and wash the blot twice for 15 min each in mPBS.

15. Transfer the blot to a clear (non-SG) container and resume at step 4 to apply another primary antibody. Alternatively, cover the blot in plastic wrap while moist and store at −80°C between pieces of cardboard held together with rubber bands.

[a] *Non-specific bands*. If non-specific antibody reactivity is seen on blots, an additional step can be added to the procedure which minimizes non-specific background. This involves first incubating the blot with secondary antibody alone at step 4, instead of applying the primary antibody, and then proceeding through the protocol (steps 8–10 and then steps 12–15). Even better results are typically obtained by performing a pre-screen using pre-immune serum or a negative control monoclonal antibody in substitution for the desired primary antibody (at step 4), followed by secondary antibody (steps 4–10 and then steps 12–15). In these cases, the pre-screening development of the blot need not include use of the ECL substrate (skip step 11). Instead, proceed directly to colorimetric (SG or DAB) 'burn-out' of the non-specific bands detected by pre-immune or secondary antibodies. After the pre-screening has been accomplished, the protocol can be resumed at step 4, now using the desired primary antibody; however, it is sometimes useful to repeat the pre-blocking (step 2) before addition of primary antibody. Moreover, for best results, the secondary antibody solution, which has been pre-adsorbed on the blot during pre-screening, should be reused when probing the same blot with the desired primary antibodies. This blot pre-treatment with secondary antibody can substantially reduce the background in ECL-based detection procedures.
[b] *Primary antibody sequence*. The membrane will become progressively darker with repeated cycles of antibody probing and SG substrate treatment. Most epitopes appear to withstand this problem, but some do not. We therefore empirically determine which primary antibodies produce weaker signals (either because the titre is low, the affinity is low, or the antigenic epitope is labile) and perform the immunodetections using those antibody reagents first, followed by more robust antibodies.
[c] *Overlapping bands*. The quenching of HRPase using SG or DAB substrates creates an insoluble, coloured precipitate on the membrane. The effected area (bands) on the membrane, therefore, is typically unavailable for subsequent rounds of ECL-based immunodetection. Thus, if two proteins co-migrate in SDS–PAGE, it may prove difficult to detect them using this method. For the same reason, one cannot use this method using two or more antibodies that react with different epitopes on the same protein.

2.2.2 Materials

The chromogenic HRPase substrate SG (cat. no. SK-4700), avidin–biotin complex (ABC) with HRPase (cat. no. PK-6100), biotinylated goat anti-rabbit IgG (cat. no. BA-1000), horse anti-mouse IgG (cat. no. BA2080), and goat anti-mouse IgM (cat. no. BA-2020) can be purchased from Vector Labs, Inc. (Burlingame, CA). HRPase-conjugated goat anti-rabbit IgG (cat. no. 172-1013) and goat anti-mouse IgG (cat. no. 170-6520) from Biorad, Inc. (Richmond, CA). HRPase-conjugated sheep anti-mouse Ig can be obtained from Amersham, Inc. (cat. no. NA-931) (Buckinghamshire, England). The ECL Western blotting detection reagent (cat. no. RPN2106; Amersham Inc.) is prepared according to the manufacturer's instructions. The chromogen, 3,3'-diaminobenzidine (DAB) can be purchased from Sigma, Inc. (St. Louis, MO; cat. no. D-5637) and dissolved at 0.5 mg/ml in 0.1 M Tris (pH 8.2). Immediately before use, 0.1 ml of 1% H_2O_2 is added to 10 ml of DAB solution followed by filtration through a 0.45 μm filter (Acrodisk; Millipore, Inc.). 3-Amino-9-ethylcarbasole (AEC) (Sigma, Inc.; cat. no. A-5754) is dissolved first at 25 mg per 1 ml of *N,N*-dimethyl formamide and then added to 50 ml of acetic acid buffer (0.1 M; pH 5.5) and filtered through Whatmann 3MM paper. Immediately before use, 25 μl of 30% H_2O_2 is added to 50 ml of the AEC-containing solution. The chromogen 3,3',5,5'-tetramethylbenzidene (TMB) can be purchased as a solution from Promega, Inc. (cat. no. W4121).

Primary antibodies employed for these studies include rabbit polyclonal antisera generated against synthetic peptides corresponding to sequences in the human (h) Bcl-2, hBax, hBcl-X, hMcl-1, and hBak proteins, mouse (m) Bax, or monoclonal antibodies to hBAD (12–19). Our antibodies are available commercially from various vendors (PharMingen; Chemicon; DAKO; Stratagene). Comparable antisera specific for these and other Bcl-2 family proteins are also available from a variety other vendors.

2.2.3 Solutions

Sodium acetate buffer (0.2 M, pH 5.5) is prepared by combining 150 ml of 0.2 M glacial acetic acid and 350 ml of 0.2 M sodium acetate.

Modified phosphate-buffered saline (mPBS) is 120 mM NaCl, 11.5 mM NaH_2PO_4, 30 mM K_2HPO_4 (pH 7.5) and should be prepared fresh each week.

TN-TBM solution is 10 mM Tris (pH 7.7), 150 mM NaCl, 0.01 % (v/v)* Triton X-100, 2% (w/v) bovine serum albumin (BSA) fraction V (Sigma, cat. no. A-6793), and 5% skimmed milk (Difco Labs, cat. no. 0032-17-3), with either 0.01% NaN_3 or 0.01% thimerosal.

3. Detection of Bcl-2 family proteins by flow cytometry (FACS)

Bcl-2 family proteins are intracellular antigens and thus their detection by in-direct immunofluorescence and subsequent analysis by flow cytometry (FACS) requires permeabilization of the cells using non-ionic detergents. Procedures are described here specifically for FACS analysis of Bcl-2 but are readily adapted to detection of other Bcl-2 family members by adjustment of the choice and amount of primary antibody employed. The immunodetection of Bcl-2 or other Bcl-2 family proteins can be combined with traditional cell-surface marker analysis, as well as with DNA content analysis or TUNEL assays for detecting cells with fragmented DNA, in two- or three-colour FACS assays.

3.1 Indirect immunofluorescence procedure

Protocol 5.

Method

11. Wash the cells three times with ice-cold PBS.

12. Aliquot 2.5×10^6 cells per 12 mm $\times$ 75 mm polystyrene centrifuge tube and fix for 10 min on ice in PBS containing 2% para-formaldehyde (2 ml).

13. Add 0.1 ml of 1% Triton X-100 (in PBS) and incubate on ice for 10 min.

14. Wash three times with 3 ml ice-cold PBS. [Washes consist of centrifug-ation at 1100 r.p.m. in a Sorval or IEC table-top ($\sim$350 g) with swing-ing bucket rotor for 5 minutes. Then decant quickly the supernatant and tap the tube once while holding upside down to knock-out excess fluid. This leaves $\sim$100–200 μl of PBS after each wash.]

15. Decant the supernatant and resuspend the cell pellet in the residual PBS ($\sim$100–200 ml). Add 10 μl of pre-blocking solution [PBS con-taining 10 mg/ml human IgG (Cappel, Inc.)] and incubate on ice for 10 minutes.

16. Add 1.25–10 μl of PBS containing anti-Bcl-2 or other monoclonal antibody (typically 0.25–2 μg of antibody is appropriate) and incubate on ice for 30 min.

As controls, the same amount of a control immunoglobulin which is matched for species (mouse, rat, hamster), isotype (IgG, IgM), and subclass (IgG1, IgG2a, IgG2b) is added in a similar manner to a duplicate cell sample.

17. Wash the cells with 3 ml of ice-cold PBS, three times.

18. After the last wash, decant the supernatant and resuspend the cell pellet in residual PBS ($\sim$100–200 μl).

19. Add secondary antibody:
    (a) 20 μl of PE-conjugated rat anti-mouse IgG1 (6.25 μg/ml) with 2% gelatin and 0.1% sodium azide (Becton Dickinson) in PBS.
    or
    (b) 100 μl of 1:100 dilution in PBS of FITC-conjugated goat anti-hamster IgG (H+L) at 1.4 mg/ml (Jackson ImmunoResearch Laboratories, Inc., West Grove, PA.).

10. Following incubation with the secondary antibody on ice for 30 min, wash the cells again with 3 ml of ice-cold PBS, three times.

11. Fix the cells in 0.5 ml of 1% formaldehyde in PBS (use methanol-free formaldehyde: purchased from Polysciences, Inc., Warrington, PA, as a 10% formaldehyde ultrapure EM grade).

12. Analyse within a few days by FACS, keeping at 4°C and protected from light. (Actually, when fixed in formaldehyde and stored in the dark, these can be analysed up to 1 month later). There is no need to wash out formaldehyde before FACS analysis.

3.2 Antibodies

1. *Bcl-2 monoclonals*. The most widely used monoclonal antibodies for detecting human Bcl-2 protein are Mab100 and Mab124, both mouse IgG$_1$, which are available from DAKO, Inc. (20). The 4D7 monoclonal (21), available from PharMingen, Inc., is also a mouse IgG$_1$ and is suitable for FACS analysis. The 6C8 antibody is a hamster IgG$_{2a}$ (22) (PharMingen, Inc.). For FACS-based detection of mouse Bcl-2 protein, we have used the 3F11 monoclonal with success (23, 24), a hamster IgG$_{2a}$ (PharMingen, Inc.).

2. *Control antibodies*. As controls, the same amount of mouse IgG$_1$ (in the case of DAKO antibody or 4D7 antibody) or hamster IgG$_{2a}$ (in the case of 6C8) are added to duplicate cell samples.

   - mouse IgG$_1$ negative control (Dako, Inc., cat. no. 931)
   - hamster IgG isotype standard (Pharmingen, Inc., cat. no. 11091D)

3. *Primary antibody optimization*. For optimal results, the amount of primary antibody employed should be titered for each type of cell undergoing FACS analysis. Using too much can reduce the signal/noise ratio. For example, based on titration experiments performed using leukaemic B cells from patients with chronic lymphocytic leukaemia (B-CLL), we have determined that 0.32–0.65 μg per sample of anti-Bcl-2 antibody Mab100 (DAKO, Inc.) is optimal for FACS quantification of Bcl-2 protein levels, whereas 0.32 μg was optimal for the anti-Bcl-2 antibody 4D7 (PharMingen, Inc.). Note that these concentrations are 5–10 times less than those recommended by the manufacturers.

4. Northern blot analysis of *BCL-2* mRNA

Reproducible detection of *BCL-2* mRNA can be challenging, particularly in human tissues and cells, because the mRNA is subject to several alternative splicing events, differential termination/polyadenylation site selection, and variable transcription initiation site usage. This results in several mRNA species, most of which contain an intact open reading frame for the Bcl-2 protein, but which vary in length and thus fail to form a tight band upon agarose gel electrophoresis and subsequent analysis by Northern blotting. Partly for this reason, as well as because *BCL-2* is a relatively low abundance mRNA in many tissues, it is often necessary to use polyA-selected mRNA.

In humans, *BCL-2* transcripts vary in size from 5.5 to 8.5 kb. In addition, shorter transcripts of 3.0–3.5 kb can also be seen which contain an open reading frame for a shorter protein, Bcl-2-β and which arise through an alternative splicing mechanism (25). The Bcl-2-β protein, however, is rarely if ever seen in cells, suggesting either that it is an unstable protein or that these shorter transcripts are inefficiently translated. In mice, *bcl-2* mRNAs are generally ~7.5 kb (full-length Bcl-2) and ~2.5 kb (Bcl-2-β) (26).

The 5′ portions of the *BCL-2* gene are GC-rich, and thus hybridization probes derived from this portion of the gene or from cDNAs corresponding to this region are difficult to work with. Background is typically high and cross-hybridization to the 28S rRNA band is problematic. Probes derived from the 3′-untranslated region (3′-UTR) of *BCL-2* cDNAs or the corresponding exon from the *BCL-2* gene are preferred and generally give clean results under standard high stringency conditions. In fact, using ^{32}P-labelled hybridization probes from the 3′-UTR, it is often possible to hybridize at usual conditions of stringency in the presence of 50% formaldehyde at 42°C, and then to wash at relatively low stringency. This results in Northern blot results that are sufficiently free of background but in which the *BCL-2* bands are stronger, thereby improving signal to noise ratios.

4.1 General guidelines for RNA blotting

The methods for preparation of RNA, gel electrophoresis, and transfer to nylon membranes can be found in any basic textbook on molecular biological methods. In general, however, we obtain best results using RNA prepared by the guanidinium thiocyanate–phenol–chloroform extraction method of Chomczynski and Sacchi (27). We also prefer formaldehyde gels as opposed to glyoxyl and have enjoyed good success with Gene Screen nylon membranes (DuPont, Inc.), though certainly nylon membranes made by other manufacturers should be adequate. The transfer is done in 10 × SSC. A brief alkaline treatment (5–15 min) prior to transfer can improve the transfer efficiency but also reduces the signal in many cases and thus we do not routinely employ it. To encourage transfer of the relatively large *BCL-2* mRNAs, we typically use a 0.8% agarose gel and allow the electrophoresis to proceed until the bromophenol blue marker has run off the gel. This brings the *BCL-2* mRNA down towards the centre of the gel and probably ensures better subsequent transfer. We also transfer for a full 24 h and change the blotting papers several times during the transfer, replenishing the transfer solution as needed. A T75 cm^2 tissue-culture flask, filled with water, and placed on its side makes a good weight for placement on top of the paper towels during transfers. We do not have experience with the recently described downward capillary transfer technique (28), which has been claimed to improve transfer efficiencies. In the past, we have added ethidium bromide (7.5 µg) directly to the RNA sample during loading into the gel, but recent reports that ethidium impairs transfer suggest that this may not be a good practice (29). Adding the ethidium directly, however, allows one to assess the integrity of the RNA and requires no destaining prior to photographing the gel. We do, however, limit the exposure of the ethidium-stained RNA gels to < 3 secs during the photography, to minimize damage. After transfer in 10 × SSC, the membrane is rinsed briefly in 2 × SSC, placed on 3M paper wetted in 2 × SSC, and UV cross-linked (1200 joules), then dried on 3M paper and baked in a vacuum oven at 80 °C for 1 h.

4.2 Pre-hybridization and hybridization procedures

Use RNase precautions and RNase-free solutions prepared using diethyl pyrocarbonate (DEP)-treated dH₂O.

Protocol 6.

Method

11. Float the dry membrane, RNA side up, on DEP-treated dH$_2$O. When completely wetted, submerge the filter.

12. Place into a heat-sealable plastic bag.

13. Pipette 10 ml of pre-hybridization/hybridization solution into the bag.

14. Heat-seal the bag, then cut a corner off, and careful expel all bubbles. Seal.

15. Pre-hybridize submerged in a 42 °C water bath with gentle shaking for 6 h up to overnight. Occasionally, massage the bag to ensure uniform distribution of solution.

16. Cut a corner of the bag and expel the pre-hybridization solution.

17. Pipette into the bag 10 ml of pre-warmed (42 °C) hybridization solution containing 10^6 cpm per ml of ^{32}P-labelled DNA probe.

18. Carefully expel bubbles, catching any radioactive fluid on paper towels. Heat-seal the bag and test for leaks.

19. Incubate in the 42°C bath overnight with gentle shaking. Occasionally massage the bag to mix the solution.

10. Expel radioactive hybridization solution into an appropriate container. Remove the blot and wash for 20 min at room temperature in 2 × SSC/0.1% SDS (200–250 ml per blot) with shaking.

Protocol 6. *Continued*

11. Wash at 68°C in pre-warmed 2 × SSC/0.1 % SDS for 15 min with vigorous agitation.

12. Rinse briefly (3–5 minutes) in 2 × SSC (room temperature) to remove excess SDS.

13. Dry the membrane briefly, RNA side up, on 3M paper but leave slightly damp.

14. Cover in plastic wrap and expose to X-ray film using intensifying screens at –80°C.

15. If the background is high, re-wet the membrane in 2 × SSC and wash at 68°C for 15–30 min in 1 × SSC/0.1% SDS. Expose to film. If the background is still high, try successive washes (with exposures to film in between) using 0.5 × and then 0.1 × SSC with 0.1 % SDS.

4.3 Solutions

All ingredients are made from RNase-free solutions. Refer to any standard textbook on molecular biology for details on the preparation of the stock solutions required (30). Sodium–heparin is purchased from Sigma Chemicals, Inc., dissolved in DEP-treated water, and stored in 0.5 ml aliquots at –20°C.

Make the pre-hybridization/hybridization solution fresh each time, preparing enough solution for both the pre-hybridization and hybridization. After pre-warming to 42°C, remove half for hybridization, and store at 4°C until ready to perform the hybridization (if doing overnight pre-hybridization), then pre-warm to 42°C again. The denatured DNA should be added to both the pre-hybridization and hybridization solution just before use.

Per 50 ml

Formamide (deionized)	25 ml
1 M Tris–base (pH 7.4)	2.5 ml
5 M NaCl	10 ml
50% dextran sulfate	10 ml
10% SDS	5 ml
50 × Denhardt's solution	1 ml
50 mg/ml Na–heparin	0.5 ml

Sheared or sonicated salmon sperm DNA at 10 mg/ml in TE buffer is boiled for 10 min, quick-cooled on ice for 5 min, and 0.25 ml added per 10 ml.

DNA probes are radiolabelled by the random primer method, purified by centrifugation through Sephadex G50, and boiled in TE buffer for 10 min, then cooled on ice for 5 min before addition to hybridization buffer. The unused probe can be stored at $-20°C$ and used for up to 2 weeks in many cases. The specific activity of probes should be $\sim 10^9$ cpm/μg.

5. Reverse transcriptase-polymerase chain reaction (RT-PCR)

An alternative to Northern blotting for detecting and estimating relative amounts of *BCL-2* mRNA is RT-PCR. If performed in a semi-quantitative fashion, this technique can allow for monitoring of fluctuations in *BCL-2* mRNA levels within cells and requires far fewer cells than Northern blotting experiments. We have used the RT-PCR approach, for example, to monitor the effects of antisense oligonucleotides targeted against the human *BCL-2* mRNA, which induce an RNAse-H-like degradation of mRNAs, as well as for assessing the effects of p53 on *bcl-2* mRNA levels in murine leukaemia cells (31–33).

Total RNA can be prepared by either the guanidinium isothiocyanate procedure described above (27) or using a commercially available, streamlined version of the method involving the TRIzol™ reagent (BRL/Gibco, Inc.). In some cases, it may be advisable to check 1 μg of the RNA in a minigel (formaldehyde-containing) to assess integrity of the 28S and 18S rRNA bands, assess whether contaminating DNA is present, and verify the quantification results (i.e. the starting amount is approximately the same for all samples).

The procedure described below involves use of a recombinant version of Moloney Virus-derived RTase, SuperScript™ (BRL/Gibco, Inc.) for the cDNA synthesis and Taq DNA polymerase (Perkin–Elmer, Inc.) for the PCR, but other RTases and heat-stable DNA polymerases may also be suitable.

It is important to also perform RT-PCR for the same sample of RNA using a control set of primers. For mouse, we typically target β_2-microglobulin mRNA, whereas for human GADPH is used. The primer sequences for amplification of these control mRNAs are also provided. Of course, a variety of other control mRNAs could be equally, or even more, appropriate, depending on the particular cells of interest. The value of the control RT-PCR is that it permits the *bcl-2* results to be normalized relative to a control that should (in most circumstances) not vary.

The procedures described here are designed specifically for detection of *bcl-2* RNAs but can be adapted through use of alternative amplification primers for any of the Bcl-2 family proteins.

5.1 cDNA synthesis

Protocol 7.

Method

1. Pipette 5 μg of total RNA into a 0.5 ml microcentrifuge tube and bring the total volume to 24.5 μl using DEP-treated water.

2. Briefly heat (95°C, 3 min) and then quickly cool on ice.

3. Add the following reagents.

   5 × first strand buffer (GIBCO-BRL), 20 μl

   RNAsin (Promega) 20 units, 0.5 μl

   random hexamers (62.5 A_{260} units/ml), 4 μl

   dNTPs (2.5 mM each), 40 μl

   0.1 M DTT, 10 μl

   SuperScript™ reverse transcriptase (GIBCO-BRL) (200 units), 1 μl

4. Incubate at 37°C for 1 h.

5. Heat at 95°C for 5 min to terminate reactions.

5.2 PCR amplification

Protocol 8.

Method

1. To a 0.5 ml microcentrifuge, add 8–10 μl of the cDNA reaction mixture from above.

2. Add the following reagents.

   DEP-treated water, 75–77 μl

   10 × reaction buffer (Perkin–Elmer Cetus),[a] 10 μl

   forward primer (25 pmol/ml), 2 μl

   reverse primer (25 pmol/ml), 2 μl

   dNTPs (2.5 mM each), 2.4 μl

   Taq polymerase (5 units/ml), 0.5 μl

   Final concentration is 25 mM TAPS [pH 9.3], 50 mM KCl, 2 mM $MgCl_2$, 1 mM β-mercaptoethanol.

3. Overlay with 50 μl mineral oil and heat to 94°C for 7 min (alternatively, Taq can be omitted until after the heat denaturation, then pipetted under the mineral oil layer into the aqueous solution).

172

4. Amplify *bcl-2* cDNA using the following PCR cycling conditions:

   Mouse bcl-2:

   94°C, 30 sec

   57°C, 30 sec

   72°C, 3 min (this should be 8 min for the last cycle)

   Human BCL-2:

   94°C for 1 min

   72°C for 1 min, where the 72°C step is lengthened by 8 sec per cycle

   (A final extension at 72°C is performed for 10 min after the last cycle)
   The number of cycles for obtaining results within the linear portion of the
   reaction varies between cells and tissue types. In general, 25–30 cycles
   has been appropriate for mouse and 25–20 for human, but a range of
   cycle numbers should be examined in pilot experiments, e.g. 15, 20, 25,
   30, 35 cycles.
   The two cycle profiles for mouse and human *BCL-2* were independently
   arrived at by separate workers in the laboratory and have proved to be
   empirically successful in our hands. No efforts have been made to test the
   interconvertibility of these cycling profiles for mouse and human *BCL-2*.

5.3 Southern blot analysis of PCR products

Protocol 9.

Method

1. Prepare a 3% agarose gel using 3:1 ratio of NuSieve agarose to
   agarose (FMC Bioproducts, Inc.; Rockland, ME).

2. Mix 20 ml of the RT-PCR reaction product with 10 μl of TAE electrophoresis buffer. Add 6 μl of 6 × loading dye #3 (30) and heat at 68°C
   for 10 min, followed by quick cooling on ice for 5 min.

3. Load samples into the wells of the gel and perform electrophoresis at
   ~30 volts/cm until the bromophenol blue marker has run ~2/3 to 3/4
   the length of the gel.

4. If using unlabelled DNA markers, briefly stain gel with ethidium and
   photograph with UV transillumination (add 50 μl of 10 mg/ml ethidium
   bromide to ~500 ml of TAE, using the old electrophoresis buffer). The
   expected PCR product for the human is 318 bp; and for the mouse
   575 bp.

5. Transfer the size-fractionated DNA to GeneScreen Plus™ nylon
   membranes (New England Nuclear/DuPont, Inc.) using 0.4 N NaOH as

Protocol 9. *Continued*

the transfer buffer. Alternatively a solution of 0.1 N NaCl and 0.1 N NaOH can be used to transfer to Zeta-Probe membranes (Biorad, Inc.).

6. If using Gene Screen Plus membranes, no baking or UV cross-linking is required. The filters are merely, rinsed in 2 × SSC for 5–10 min to neutralize NaOH, before proceeding to pre-hybridization.

For Zeta-Probe or similar nylon membranes, neutralize the filter for 10 min in 2 × SSC, and UV cross-link (1200 joules).

7. For the mouse *bcl-2*, we hybridize overnight at 57°C with ~10 pmoles of ^{32}P-end-labelled internal oligonucleotide probe in 5 × SSPE (1 × SSPE=0.15 M NaCl/0.01 M NaH$_2$PO$_4$, pH 7.0/1 mM EDTA), 0.6 % SDS, 50 μg/ml heat-denatured salmon sperm DNA. Washes of the filter are then performed in 5 × SSPE and 0.1 % SDS at 57°C for 20 min (twice).

8. For the human *BCL-2*, the procedure we have used entails hybridization with 10 pmoles of ^{32}P-end-labelled oligonucleotide probe in 5 × SSC containing 0.1% SDS, 5 × Denhardt's (30), and 100 μg/ml salmon sperm overnight at 50°C.

Washes are performed with 5 × SSC and 0.1% SDS twice for 10 min at room temperature, followed by exposure to X-ray film [Kodak XAR film (Eastman-Kodak, Rochester, NY)] overnight at –80°C with intensifying screens.

These two alternative approaches have not been compared side-by-side to determine whether one is superior to the other, but both have reproducibly yielded excellent results in our hands.

5.4 Primers and probes for PCR

5.4.1. Mouse *bcl-2*

Forward primer: 5′-TGCACCTGAGCGCCTTCAC-3′

Reverse primer: 5′-TAGCTGATTCGACCATTTGCCTGA-3′

Oligonucleotide internal probe: 5′-CCAGGAGAAATCAAACAAAGG-3′(expected size of the PCR product is 575 bp) {Oligonucleotides are end-labelled with [^{32}P]γATP using T4 polynucleotide kinase. Unincorporated [^{32}P]ATP is separated from the probe using either ethanol precipitations in the presence of carrier tRNA, DEAE–cellulose minicolumns (Whatman DE-52), or molecular sieve chromatography [centrifuge through Bio-spin 6 minicolumns (BIORAD, Inc.) (exclusion limit 6000 daltons or ~5 bases)].}

5.4.2 Human *BCL-2* primers and probes

Forward primer: 5'-CGA CGA CTT CTC CCG CCG CTA CCG C-3'

Reverse primer: 5'-CCG CAT GCT GGG GCC GTA CAG TTC C-3'

Oligonucleotide internal probe: 5'-GGC GAT GTT GTC CAC CAG GGG
 CGA C-3' { Oligonucleotides are end-labelled with [^{32}P]γATP using T4
 polynucleotide kinase. Unincorporated [^{32}P]ATP is separated from the
 probe using either ethanol precipitations in the presence of carrier tRNA,
 DEAE–cellulose minicolumns (Whatman DE-52), or molecular sieve
 chromatography [centrifuge through Bio-spin 6 minicolumns (BIORAD,
 Inc.) (exclusion limit 6000 daltons or ~5 bases)].}

5.4.3 Mouse β$_2$-microglobulin (for internal control)

Forward primer: 5'-ATGGCTCGCTCGGTGACCCTAG-3'

Reverse primer: 5'-TCATGATGCTTGATCACATGTCTCG-3'

Oligonucleotide internal probe: 5'-GCTACGTAACACAGTTCCAC-3'
 (expected size of the PCR product is 373 bp)

5.4.4 Human GADPH

Forward primer: 5'-CCA CCC ATG GCA AAT TCC ATG GCA-3'

Reverse primer: 5'-TCT AGA CGG CAG GTC AGG TCC ACC-3'

Hybridization oligonucleotide probe: 5'-CAA CAC AGA CCC ACC CAG
 AGC CCT CCT GCC CTC CTT CCG CGG GGG C-3' (expected PCR
 product is 598 bp in length)
 For GAPDH, the thermocycler profile is as follows:

Heat samples at 94°C for 7 min.

Cycle=94°C for 1 min, 55°C for 1 min, 72°C for 1 min (The numbers of cycles
 of PCR must be empirically adjusted to find the linear-range for your cells
 (30, 25, 20 and 15 cycles.)

Perform a final extension at 72°C for 10 min prior to termination reactions

Analyse one-fifth of the PCR product

6. Immunohistochemical detection of Bcl-2 family proteins in tissues

The approach we prefer for immunohistochemical detection of Bcl-2 protein
is to employ sections derived from paraffin-embedded tissues (*Figure 1*). This
yields morphology which is far superior to cryostat sections and permits use of

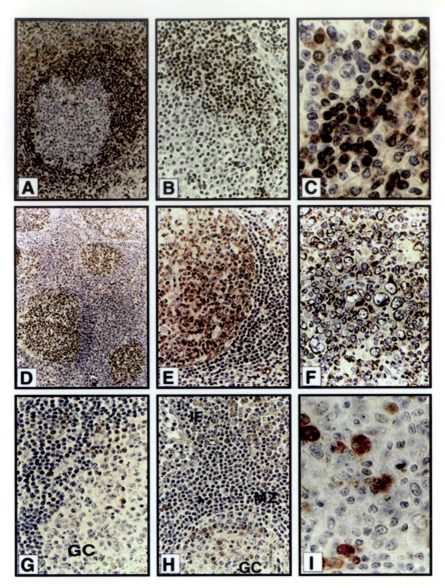

Figure 1. Immunohistochemical analysis of Bcl-2 family protein in tissue sections. Examples are provided of immunohistochemical analysis of Bcl-2 (A–C), Mcl-1 (D–F), and Bcl-X (G–I) in lymph nodes. Photomicrographs are presented at low (first column), medium (second column), and high (third column) magnification. Immunodetection was accomplished by a diaminobenzidine method (brown) and nuclei were counterstained with haematoxylin (blue). The tissue sections highlight the different patterns of Bcl-2 family protein expression in the germinal centre (GC) lymphocytes of the lymphoid follicles and the surrounding mantle zone region and interfollicular zones. (A–C) Bcl-2 staining is found primarily in the mantle zone lymphocytes surrounding the germinal centres and in occassional cells in the interfollicular regions. (D–F) Mcl-1 immnostaining is strongest in germinal centre lymphocytes. (G–I) Bcl-X staining is found primarily in plasma cells and occassional activated lymphocytes.

archival clinical material. A critical aspect of the procedure is either fixation of the tissues in an acidic fixative or, if fixed in a neutral solution, subsequent treatment with an acidic solution. The lower pH somehow reveals the epitopes that are recognized by all of the currently available commercial antibodies, thus markedly improving sensitivity. This same approach has also yielded superior results for antibodies raised against other members of the Bcl-2 protein family, such as Bcl-X, Mcl-1, and Bax (12–19). Several colorimetric or fluorescence-based detection systems can theoretically be employed for visualization of the antibodies, but most of our experience is with horseradish peroxidase (HRPase)-based colorimetric detection methods.

6.1 Procedures for tissue preparation, embedding, and sectioning

6.1.1 Fixation

Tissues should be quickly fixed with acidic based solutions such as Bouin's solution (Sigma Chemicals Inc.; cat. no. HT10-1-128) by immersion or if possible from animals, by perfusion with 2% paraformaldehyde in PBS or 4% neutral-buffered formalin and subsequent post-fixation in Bouin's solution. (Bouin's is an acidic fixative and thus will cause DNA nicking. Do not use Bouin's if planning to do *in situ* detection of DNA fragmentation.)

However, very satisfactory immunostaining can often be achieved using routine archival material that was fixed in 10% buffered formalin and then embedded in paraffin. Thus, if such material is available, it is worth a try.

Protocol 10. Immersion method

Method

1. If tissues are derived from animals that were not perfused or from human autopsy specimens, put excised tissues/organs into fixative for 3–5 h (Bouin's) or overnight (10% neutral-buffered formalin) in intact form, then cut in half to allow for better infiltration of fixative if a large organ (e.g. liver, heart, muscle, spleen). Cutting is generally not required for mouse lymph nodes or thymus because the tissues are small. [For brain tissue, fix intact overnight (even if using Bouin's), then cut the next day.]

2. Let the tissues fix for an additional 3–5 days, depending on the size of the tissues (e.g. mouse/rat for 3 days vs. human autopsy for 5 days). Typically, the fixation is done in plastic beakers/jars.

   If tissues are derived from perfused animals (see below), the post-fixation time can be reduced to 1–2 days and the formalin is only 4% instead of the usual 10%. Post-fixations in Bouin's solution are at usual full strength.

Protocol 10. *Continued*

3. Place tissues into cassettes for paraffin embedding. Label with a 'superfrost' marking pen which is resistant to alcohols and xylenes. It may be necessary to cut the tissues smaller with a scalpel to get them to fit. However, do not let the tissue pieces dry out at any point. Also, orient the tissues in the direction you will want them cut, then instruct the histotechnician to organize the tissues in the paraffin molds in the same fashion as they appear in the cassettes. (The bottom of the cassette is the cutting surface.)

4. Submerge the cassettes for a few hours in 50% EtOH/PBS and then proceed to dehydrations using an automatic tissue processor.[a] Alternatively, if fixed tissues will not go directly to embedding, dump out the fixative and replace with PBS. Change the PBS solution twice over the next day or so and store the fixed tissues in PBS at 4°C. These fixed tissues should be good for at least a few weeks, but add azide if keeping longer or if shipping unembedded tissues at room temperature.

[a] Our tissue processor is a Shandon, Inc. (model Citadel). This must be optimized for your own processor, but the conditions we use are: (a) 50% ethanol/PBS for 1 h; (b) 70% ethanol/water for 1 h; (c) 95% ethanol for 1 h; (d) 95% ethanol for 1 h again; (e) 100% ethanol for 1 h; (f) 100% ethanol for 1 h again; (g) xylene or xylene substitute (Hemo-De from Fisher, cat no. 15-182-507A or Histosol from National Diagnostics, cat. no. HS-100) for 1 h; (h) repeat xylene or xylene substitute for 1 h; (i) paraffin at 57–61°C for 1.5 h; (j) repeat paraffin at 57–61°C for 1.5 h. Note that isopropanol can be substituted for ethanol in all of these steps.

Protocol 11. Perfusion method

For paraffin immunocytochemistry, the optimal solution composition should be either: 4% formaldehyde, buffered with PBS to pH 7.4–7.6 (4% formaldehyde/PBS) or 2% paraformaldehyde in PBS. The approximate volume required for perfusion by cardiac puncture is 100 ml per mouse and 200–300 ml per rat. An i.v. infusion system with a volving system and manometer is required, equipped with appropriate tubing and a 16–18 gauge needle attached for cardiac punctures.

Method

1. Animals are first placed into deep anaesthesia, usually using Nembutal.

2. Lift the skin over the mid-epigastric region and make an incision with scissors. Spread open the hole with the scissor blades and cut horizontally to reveal the diaphragmatic muscle. Carefully cut the diaphragm

away from its attachments. Cut the ribs and lift the sternum, revealing the heart. Insert the needle, hooked to the infusion bottle containing PBS without fixative into the left ventricle. Immediately cut a small hole in the right atrium or ventricle with scissors to relieve pressure. Start the flow of PBS into the left ventricle with pressure set at 100–120 mmHg. (The animal should be in a pan to catch the fluid that will drain from the right side of the heart.) Continue infusion of PBS at 100–120 mmHg until essentially all blood has been removed from the vasculature: i.e. the flow from the right ventricle is clear PBS and not blood.

3. Switch to the fixative infusion bottle and infuse fixative until the body of the animal becomes stiff (about 10 min for mice and 15–20 min for rats).

4. Discontinue the infusion, and excise the tissues of interest.

5. Place the tissue into post-fixative which is either the same formalin solution used for perfusion or Bouin's solution.

6. If sufficient material is available, it is often advisable to split it into two portions. Post-fix one of the pieces in Bouin's solution, which is generally the better fixative for Bcl-2 protein family immunocyto-chemistry (Sigma Dgn.; cat. no. HT10-1-128). Post-fix the second piece in 4% buffered formalin for immunostaining with other antibodies or for DNA fragmentation analysis by the TdT end-labelling method (34). (Bouin's solution fixation results in non-specific DNA breaks.)

[a] Tissue processing and paraffin embedding is accomplished as in Protocol 10 above.

6.1.2 Paraffin embedding

Do not overheat or overdehydrate the samples. Keep the temperature of the paraffin no higher than 65°C. We use the 56–60°C melting-temperature paraffin from Surgipath, Inc. (cat. no. E1-600) or equivalent. (Note: The two pieces fixed differently can be embedded in the same paraffin block under the same patient/experiment number. For example, we have adopted the convention of placing the Bouin's-fixed piece at the top and the formalin-fixed piece at the bottom of the slide.)

6.1.3 Sectioning

Paraffin sections 3–5 μm are prepared and mounted on poly-L-lysine coated slides or Fisher-Superfrost-plus slides (generally 50–200 slides can be obtained per block, but of course it depends on the size and thickness of the tissue specimen).

If the paraffin samples are small, 2 or 3 separate patient samples or experimental animal samples can be placed adjacent to each other on the same glass slide so as to minimize the labour of immunostaining.

6.1.4 Solutions

Modified PBS.

NaCl 7 g/litre

NaH$_2$PO$_4$ ($\times$ H$_2$0) 1.38 g/litre

K$_2$HPO$_4$/anhydrous 5.44 g/litre

adjust pH to 7.4–7.6

4% Neutral-buffered formalin/PBS.

add 108 ml of 37% formaldehyde

bring volume to 1 litre using modified PBS

readjust pH to 7.4–7.6 with 10 N NaOH

10% Neutral-buffered formalin/PBS

first make 2 $\times$ PBS solution by combining the salts described above into 0.5 litres instead of 1 litre. Then mix:

500 ml of 2 $\times$ PBS

270 ml of 37% formaldehyde

230 ml of dH$_2$O

adjust pH to 7.4–7.6. with 10 N NaOH

6.2 Pre-staining sample preparation

Protocol 12. Deparafinization procedure

Method

1. Pre-warm the slides in an oven at 55°C for 1–3 days. This enforces attachment of the tissue to coated slides. (If you receive your slides from an outside source, check whether this has already been done.) The slides should be placed into a Wheaton glass slide rack for this procedure.

Any oven can be used for this. A standard vacuum oven found in most molecular biology laboratories is suitable. We typically hook to house vacuum line and turn on the vacuum, which helps to keep the temperature constant (~15 mmHg).

---

2. Remove the slides from the oven. Then pre-warm the oven to 65°C before reinserting.

3. Melt the paraffin at 65°C for 20 min (the drops of paraffin should be seen at the bottom of the slides).

4. Set up three separate glass containers (Wheaton/Fisher, cat. no. 08-812) containing xylene or xylene substitute (such as Hemo-D from Fisher Scientific, cat. no, 15-182-507A) and transfer the slides while still in the glass slide rack sequentially to each jar for 2 min each. Agitate the slides up and down a few times during each transfer before submerging for 2 min, then agitate up and down before transferring to the next container of xylene. Shake off drops before transferring to the next solution.

5. Immerse slides in 100% ethanol for 15 sec.

6. Place into 3% H_2O_2 in methanol for 30–45 min.

7. Next, transfer to 95% ethanol for 20 sec.

8. Then transfer to 70% ethanol for 20 sec.

9. dH_2O for 1 min.

From this point on, keep the slides moist. DO NOT let them dry out.

Protocol 13. Antigen retrieval (acid treatment)

Microwaving in acidic solutions is used to enhance antigen reactivity.[a]

Method

1. Transfer slides to four polyethylene Coplin 'jars' (Fisher Scientific, Inc.; cat. no. 08-815-10) and place these jars at the four corners of a plastic box. Each container should have 10 glass slides. Insert 'blank' glass slides if your experiment involves fewer than 40 slides.

2. Add acidic solution (pH 5.5–6.0) into the containers until the slides are just covered. Also put tap water into the plastic box until the bottom is covered.

3. If your microwave does not have a rotating stand, after each pulse of microwaves change the positions of the flasks (it is convenient to simply move the container clockwise or counter-clockwise so that each of the coplin jars will have occupied all four corners of a square by the end of the procedure). Also gently agitate each container after each heating to ensure uniform heating of the fluid

Protocol 13. *Continued*

4. For our oven (General Electric; model JE1453H-001), we heat as follows:

Power setting	Time (min)	Desired temperature (°C)
7	1	
6	1	
5	2	85
6	1	
5	0.5	90–95

In general, however, the goal is to bring the solution in which the slides are submerged close to a boil (~90–95°C) without boiling. (If boiling does occur for a few seconds, however, it is usually not deleterious.) Once brought to a temperature of 90–95°C, the sample need *not* be held at this temperature. Simply let cool it to room temperature.

5. Let the containers cool at room temperature for 15 min.

6. Wash twice in PBS (pH 7.4–7.6) for 5 min each.

[a] Microwaving should be performed at a slightly acidic pH. The dH_2O at our institution has a pH of ~6.0 and we therefore use it directly. Alternatives include 10 mM citrate buffer at pH 6.0 or 50 mM acetate buffer at pH 5.5. These conditions should be optimized for your particular microwave and water source.

Protocol 14. Alternative antigen retrieval protocol

An alternative to microwaving is to treat the slides with more acidic solutions without heating. Though results are superior with microwaving, some tissue sections can detach from the glass microscope slides during microwaving (depending on how they were prepared) and therefore this alternative procedure is also provided here.

Method

1. Incubate slides in 0.02 M citrate buffer (pH 3.2) for 2 h.

2. Wash in PBS for 5 min.

6.3 Immunostaining

Protocol 15. Immunostaining procedure

Materials

- primary antibodies. Polyclonal rabbit anti-sera to Bcl-2 family proteins or mono-clonals derived from mouse or hamster
- glass dishes with removable horizontal tray that holds slides (20 slides capacity; Fisher/Wheaton, cat. no 08-812)
- closed humidity chamber; prepared from plastic boxes with plastic pipettes (10 ml) fixed at the bottom with superglue. Put tap water in the bottom of the box

- flat-tipped forceps
- reagents for avidin–biotin complex preparation (Vector Lab. ABC-Kit Standard; cat. no:PK-6100)
- secondary antibodies. Biotinylated goat anti-rabbit IgG (Vector Labs, cat no: BA-1000) or biotinylated horse anti-mouse IgG (Vector Labs, cat no: BA-2000) or bio-tinylated sheep anti-hamster IgG (Vector Labs, cat no: BA-9100)

Method

This procedure is optimized for final detection of antibodies using an avidin–biotin complex reagent and diaminobenzidine colorimetric reaction involving horseradish peroxidase. Other detection methods may require some modifications. Perform all of the following steps in a humidity chamber at room temperature.

Do not let the slides dry at any point. Keep covered with sufficient volume of various reagents (~500 μl per slide).

1. Pre-block by incubating in slides in either 0.1 M Tris (pH 7.6–7.8) or TSK [Tris/sodium chloride/potassium chloride (pH 7.6–7.8)] with 2% BSA, 1% normal goat, and 0.05% Triton X-100 for 1 h.[a,b]

2. (a) Decant the pre-blocking solution and incubate with 0.4–0.6 ml of primary polyclonal antibody diluted typically 1:800 to 1:2000 in either 0.1 M Tris or TSK containing BSA and Triton X-100. Alternatively, 3–10 μg/ml of anti-Bcl-2 murine monoclonal antibody or 2 μg/ml of anti-Bcl-2 hamster monoclonal antibody in either 0.1 M Tris–BT or TSK-BT may be used. Incubate overnight at room temperature.[c]

   (b) A negative control slide(s) should also be prepared using pre-immune serum of the same rabbit or, alternatively, normal rabbit serum. Use isotype- and subclass-matched control monoclonals when using anti-Bcl-2 monoclonals (Dako, Inc., Santa Barbara, CA). If no negative control antibodies are available, perform immunostaining using only the secondary antibody, without a primary antibody.

   (c) An excellent control entails pre-adsorbing the antibody with excess antigen (peptide or recombinant Bcl-2 protein). For peptides, we typically add 10 μl of antiserum to 1 ml of TSK-BT and 10 μl of a solution containing the appropriate Bcl-2 peptide at 1 mg/1 ml. The antigen and antibody are incubated at 37°C for 2 h before use.

183

Protocol 15. *Continued*

3. The next day, wash slides three times in PBS for 5 min each using plastic Coplin jars. Agitate gently to wash. Do not put slides with different antibodies (or even different dilutions of antibodies) together in the same washing container.

4. Incubate with ~0.5 ml of secondary antibody for 1 h in a humidity chamber.

   (a) For polyclonal antisera: 3–5 µg/ml of biotinylated antibody from goat anti-rabbit immunoglobulins (Vector Lab.; cat. no. BA-1000) in 0.1 M Tris–BT or TSK-BT, with 0.5% normal serum.[d] (Usually 5 µg/ml is used.)

   (b) For mouse monoclonals: 2.7 µg/ml of biotinylated horse anti-mouse IgG (Vector Labs, Inc.; cat. no. BA-2000) in either 0.1 M Tris–BT or TSK-BT with 0.5% normal serum.[d]

   (c) For hamster monoclonals: 2.5–2.7 µg/ml of biotinylated sheep anti-hamster IgG (Vector Labs, Inc.; cat. no. BA-9100) in either 0.1 M Tris–BT or TSK-BT with 0.5% normal human serum.

5. Wash three times for 5 min each in PBS. (Keep slides separated into different Coplin jars if they received different secondary antibodies and/or different dilutions).

6. Incubate slides in a humidity chamber for 45 min in ~0.5 ml each of avidin–biotin complex reagent containing horseradish peroxidase (HRPase) (Vector Labs.; Vectastain Kit, cat. no. PK-6100). [This solution must be prepared in 0.1 M Tris (pH 7.4–7.8) 1 h before use!]

7. Wash with PBS three times for 5 min each. (All slides can be washed together at this point, regardless of primary or secondary antibodies or the dilutions of antibodies, with the exception that we generally take the additional precaution of washing separately any negative control slides, e.g. those that received pre-immune serum or peptide-adsorbed antibodies, etc.)

8. Develop the reaction product for 10–20 minutes in a humidity chamber by covering slides in DAB solution containing H_2O_2, as described below. (This solution must be prepared freshly each time!)

9. Wash slides in dH_2O using plastic Coplin jars three times for 5 min each.

[a] We generally use 0.1 Tris (pH 7.6–7.8) but if high background is a problem, try TSK. The high salt concentration eliminates much non-specific immunostaining.
[b] The use of goat serum assumes that the secondary antibody employed is derived from goat. If a different species is employed for secondary antibody production (horse, sheep, donkey, rabbit), then 1% of normal serum from that species should be used instead of normal goat serum.
[c] If non-specific immunostaining is seen with polyclonal antisera raised against synthetic peptides, pre-adsorption of the antibody with the same protein used as a carrier (ovalbumin, KLH) can be helpful. The optimal amount of murine monoclonal depends on the quality of fixation, etc., but usually ~10 µg/ml for Mab124 and ~2 µg/ml for 407.
[d] Include 0.05–0.1% normal serum from the same species that the tissue sections are derived from, e.g. if immunostaining human tissue then add human serum.

6.4 Counterstaining and mounting

Protocol 16.

Perform all steps using glass slide racks. Counterstaining with methyl green may be preferred to haematoxylin when studying lymphocytes because of the scant cytoplasm (high nuclei/cytosolic ratio). However, methyl green will fade with time, especially if not kept in the dark. Generally, it is wise to photograph such slides within 2 weeks.

Method

1. Rinse slides in tap water for 1 min.

2. Submerge slides in haematoxylin counterstaining solution (Mayer's Hematoxylin; Sigma, Inc., cat. no. MH5128) for 20 sec to as much as 5–15 min. (The optimal time depends on the source of haematoxylin and how fresh the solution is; it will take longer after many uses.) Alternatively, methly green can be used for counterstaining.[a]

3. Rinse slides in tap water for 1 min.

4. Place slides, still in the slide racks, into ~200 ml of tap water containing 5–7 ml of NH_4OH (28–30%; Aldrich Chemicals, Inc., cat. no. 1336-21-6) for 15–20 sec.

5. Rinse in tap water again.

6. Rinse in dH_2O for 1 min.

7. Dehydrate samples by submerging for 15–30 sec each in sequence into 70% EtOH, then 95% EtOH, then 100% EtOH, followed by three treatments with xylene or a xylene substitute.

8. Place a drop of acrytol mounting media (Surgipath; cat. no. MM-160) on to the slide over the tissue or cells, and apply a coverslip.

9. Let dry at room temperature for a few hours, while lying flat with the tissue side up.

[a] (a) Counterstain in methyl green in a glass Coplin jar for 10 min at room temperature. (If the methyl green solution has be reused several times, increase the counterstaining time to 15–25 min.) (b) Wash three times in three changes of distilled water in a Coplin jar, dipping the slide 3–5 times each in the first and second washes, followed by 15 sec without agitation for the third wash. (c) Wash in three changes of 100% butanol in a Coplin jar, dipping the slide 10 times each in the first and second washes, followed by 10 sec with out agitation/dipping for the last wash. (d) Dehydrate in three changes of xylene for 2 min for each of the first two washes and for 5 min on the last wash. (e) Mount.

6.5 Solutions

1. *Modified PBS*
- sodium chloride, 7 g
- sodium phosphate ($NaH_2PO_4.H_2O$), 1.38 g
- potassium phosphate, dibasic, anhydrous (K_2HPO_4), 5.44 g
- bring volume to 1 litre with dH_2O and adjust pH to 7.6 with 10 N NaOH.

2. *0.02 M Citrate buffer*
- solution A: 0.01 M citric acid (MW 192.1), monohydrate (Sigma, C-0759), 0.192 g in 100 ml dH_2O
- solution B: 0.01 M citric acid (MW 294.1), trisodium salt (Sigma, C-8532), 0.294 g in 100 ml dH_2O
- combine 87.4 ml of A and 12.6 ml of B.
- adjust pH to 3.2 using concentrated HCl

3. *0.1 M Tris*
- Trisma base 12.1 g
- add dH_2O to ~950 ml
- adjust pH to 7.6–7.8 with HCl
- bring total volume to 1 litre

4. *TSK*
- Trisma (Sigma) base 12.1 g=100 mM final
- sodium chloride 32.0 g=550 mM final
- potassium chloride 0.8 g=10 mM final
- add dH_2O to 950 ml and adjust to pH 7.6–7.8 with HCl
- adjust total volume to 1 litre

5. *TSK-BT*
- 2 g BSA (crystalline powder)
- add TSK to 100 ml
- add 0.05 ml Triton X-100

6. *Diaminobenzidine/H_2O_2 solution*

Caution. DAB is a carcinogen. Use precautions and dispose of properly.
- dissolve diaminobenzidine in 0.1 M Tris (pH 8.1) at 0.5 mg/ml [3,3'-diaminobenzidine from Sigma powder (cat. no. D-5637) or in tablet form (D-5905)]. The final pH should be ~7.6–7.8
- add 41 µl of 30% H_2O_2 per 50 ml of DAB solution
- load into a 60 cm^3 syringe

- attach a 0.45 μm filter disk
- filter the DAB on to the glass slides
- H_2O_2 should be purchased anew every 2–4 weeks
- do not add the H_2O_2 to the DAB solution until just before applying on to the slides

7. *Methyl green counter stain*

- add 1 g methyl green dye to 200 ml 50 mM acetate buffer (pH 5.5) in a glass container
- add ~50 ml chloroform and shake vigorously
- allow layers to separate
- discard bottom (chloroform) layer
- repeat chloroform extractions until all traces of methyl violet (purple colour) have disappeared
- allow the methyl green staining solution to stand in an open flask overnight so that any residual chloroform evaporates
- if crystals remain, filter through Whatman 3MM paper
- the solution is stable for several months at room temperature
- methyl green can be reused several times. Filter through Whatman 3MM paper or equivalent

8. *Acetic acid solution*

- prepare a 0.2 M acetic acid solution using glacial acetic acid (Fisher Sci., Inc.; cat. no. A38-212). Add 11.55 ml glacial acetic acid to 988.45 ml dH_2O.
- prepare a 0.2 M sodium acetate solution. Dissolve 27.2 g sodium acetate (Sigma; cat. no. S-8625) into 1 litre dH_2O.

9. *10 mM Citric acid (pH 6.0)*

- dissolve 2.1 g citric acid (anhydrous) (MW 192) in 1 litre of dH_2O
- adjust pH to 6.0 with ~13 ml of N NaOH

The deparaffinization solutions (xylene, EtOH, citrate buffer) may be used for 2–4 weeks, depending upon the frequency of experimental procedures. The 2% H_2O_2 in methanol should be freshly prepared every two weeks. PBS should be prepared fresh each week. DAB should be made fresh each time.

7. Immunohistochemical detection of Bcl-2 protein in cultured cells

The overall procedure for immunostaining cultured cells is similar to that used for tissue sections in paraffin. The fixative solutions are more dilute and/or are used for only a short time, owing to the rapid penetration. Because most tissue culture medium contains biotin, which can cause non-specific staining when using avidin–biotin-based detection methods, an alkaline phosphatase detection system or Peroxidase Anti-Peroxidase Complex (PAP) (35) can be employed as an alternative to DAB. The sensitivity with these methods, however, is generally less than with DAB and therefore DAB should be tried first.

7.1 Procedures

7.1.1 Adherent cells

Prior to seeding the cells into 100 mm plastic culture dishes, put autoclaved microscope coverslips on the bottom of Petri dishes (four coverslips per 100 mm dish) and let the cells adhere and grow on them to the desired density.

Protocol 17.

Method

1. Place coverslips with cell side up into a plastic petri dish (~10 coverslips per dish).

2. Add 4% buffered formalin (4% formaldehyde/PBS) or Bouin's solution and incubate for 5 min.

3. Remove coverslips using forceps and dip into a container of PBS. Agitate for a few seconds in PBS.

4. Transfer for a few seconds to a second container of PBS and agitate for a few seconds.

5. Air-dry the coverslips with cell side up in a Petri containing dry Whatman filter paper.

6. Apply one or two drops of acrytol mounting medium to the bottom of the coverslip and place with cell side up on to a glass microscope slide. Let dry for 1 h or longer. If desired, each coverslip can be divided into four pieces with a diamond pencil. This permits more antibody staining conditions to be tried from a small number of cells.

7. The fixed and dried slides can be stored at 4°C for several months, or you may precede to immunostaining.

7.1.2 Suspension cells

Protocol 18.

Method

11. Take 5–10 ml cells from tissue culture or ~0.2 ml of whole blood.

12. Add an equal volume of PBS, and mix.

13. Spin down at 1500 r.p.m. for 5–10 minutes (~400 *g*).

14. Discard the supernatant.

15. Wash once with ~10 ml PBS. Then aspirate supernatant off.

16. Add 500 μl of Bouin's solution or 4% formaldehyde/PBS.

17. Incubate at room temperature for 3–5 min.

18. Spin down at 1500 r.p.m. for 10 min.

19. Aspirate off fixative and wash 3–5 times in PBS.

10. Resuspend the cells in ~0.5 ml PBS containing 0.01% NaN₃ and store at 4°C.

11. Prepare poly-L-lysine-coated slides; put one drop of solution (Sigma Chemicals, Inc.; cat. no. 1010) on the slide, smear the drop with a gloved finger tip, and let it polymerize at 37°C for 1–3 h.

12. Mix the fixed cell suspension before applying to the slides.

13. Place a drop on the coated slides and gently spread the drop with a pipette tip over an area of ~2 cm diameter. Up to three drops can be placed on the slide and spread over a larger area.

14. Dry at 37°C for a couple of hours.

15. Store in slide boxes or proceed to immunostaining procedure.

7.2 Immunostaining

Protocol 19.

Method

1. Rinse slides in PBS for 5 min.

2. Incubate for 30 min in 0.03% H_2O_2 in PBS.

3. Wash for 5 min in PBS.

4. Incubate for 5 min in PBS containing 0.1% glycine.

5. Wash in PBS for 5 min.

6. Pre-block and incubate with primary antibodies as described in *Protocol 17*, steps 1–3.

7. Incubate with secondary antibody (differs depending on whether using avidin–biotin–HRPase/DAB method described above versus alkaline phosphatase or PAP method for detection; see *Protocols 20* and *21*).

Protocol 20.   Alkaline phosphatase method

Method

1. Add ~0.5 ml of alkaline phosphatase (AP)-conjugated antibody to slides and incubate for 1 h in a humidity chamber.

AP-conjugated swine anti-rabbit IgG is used for detection of rabbit anti-bcl-2 polyclonals. Dilute 1:25 to 1:50 in 0.1 M Tris (pH 7.6–7.8) containing 2% BSA and 0.05% Triton X-100 (final conc. 10–20 μg/ml) (Dako, Inc.; cat. no. D306).

AP-conjugated goat anti-mouse IgG is used for detection of mouse monoclonals and is diluted in the same way (final conc. 32–64 μg/ml) (Dako, Inc.; cat. no. D486).

2. Wash three times in PBS (5 min each).

3. Add AP–substrate solution (NBT/BCIP) and incubate for 30 min.

4. Stop reactions by washing in PBS containing 20 mM EDTA.

5. Counterstain with methyl green and mount as described above in *Protocol 16*.

Protocol 21.   PAP method

Method

1. Add 0.5 ml of either swine anti-rabbit IgG (Dako, Inc.; cat. no. Z196) diluted 1:50 (final conc. 62 μg/ml) in 0.1 M Tris (pH 7.6–7.8) containing 2% BSA and 0.05% Triton X-100 (for detection of rabbit polyclonals), or rabbit anti-mouse IgG (Dako, Inc.; cat. no. Z259) diluted 1:50 in the same way (final conc. 60 μg/ml) for detection of mouse monoclonals.

2. Incubate for 1 h, then wash three times with PBS.

3. Add 0.5 ml per slide of 1:200 (~40 μg/ml) PAP complex (rabbit or mouse) in 0.1 M Tris–BT and incubate for 1 h.

4. Wash three times with PBS.

5. Cover slides with DAB/H_2O_2 solution (described above in Section 6.5) and incubate for 10–20 min.

6. Wash three times in dH_2O.

7. Counterstain and mount as described above in *Protocol 16*.

7.3 Solutions

NBT/BCIP (make fresh each time)

- dissolve 0.5 g NBT (nitroblue tetrazolium) in 10 ml of 70% dimethylformamide (DMF) (NBT is stored in desiccator at 4°C before use)
- dissolve 0.5 g BCIP (bromochloroindolyl phosphate) in 10 ml of 100% dimethyl formamide (DMF) (BCIP is stored in desiccator at –20°C before use)
- prepare TNM solution: 0.1 M Tris (pH 9.5), 0.1 M NaCl, 5 mM $MgCl_2$
- then, just prior to use, add 66 μl of NBT stock solution to 10 ml of TNM. Mix.
- add 33 μl of BCIP stock. Mix.
- use within 1 h

Note. The NBT and BCIP stock solutions were stable at 4°C for ~1 year.
PAP complex

- HRPase in a soluble complex with either rabbit anti-HRPase or mouse anti-HRPase can be purchased from Dako, Inc.
- dilute 1:200, final conc. 40 μg/ml, in 0.1 M Tris (pH 7.6–7.8) containing 2% BSA and 0.05% Triton X-100.

8. Production of recombinant Bcl-2 family proteins in bacteria

Biochemical and functional analysis of Bcl-2 family proteins can be accomplished using recombinant proteins produced in bacteria (*E. coli*). We have used such proteins for: (a) analysis of dimerization amongst Bcl-2 family proteins by surface plasmon resonance and other methods; (b) circular dichroism studies of protein secondary structure regulation; (c) investigations of ion channel activity of Bcl-2 family proteins; (d) microinjection into cells; (e) studies of the effects of Bcl-2 family proteins on isolated mitochondria; and (f) as immunogens for antibody generation (36–39). Procedures are described here for production of recombinant Bcl-2, Bcl-X$_L$, Bax, and Bid. In the cases of Bcl-2, Bcl-X$_L$, and Bax, the cDNAs encoding these proteins have been engineered to introduce a stop codon just prior the carboxyl-terminal ~20 amino acids which constitute hydrophobic membrane-anchoring domains in these proteins (36–38). This is necessary to obtain soluble proteins.

To facilitate the purification, these Bcl-2 family proteins are produced with either GST or His$_6$ tags fused to their N-terminus or C-terminus. In general, we have obtained better results with GST tagging of these proteins, compared with His$_6$ tagging. The GST versions of these recombinant proteins generally have better solubility and there are some advantages to glutathione–

Sepharose affinity chromatography compared with Ni-chelation resins. How-ever, the disadvantage of using GST fusion proteins is that enzymatic cleavage to remove the GST-tag can be problematic, leading to some accidental cleavage of the desired final product as well (e.g. Bid, Bax). The GST moiety can be left on the proteins for some types of assays, but not all. For example, the presence of a GST-tag at the N-terminus of Bcl-2 or Bcl-X$_L$ has been shown to preclude their interaction with some others types of proteins. How-ever, the presence of the GST-tag may also carry advantages, such as reduced toxicity to *E. coli*, which we have observed for GST–Bax compared with His$_6$–Bax, thus making it possible to produce 80–100 mg Bax fusion protein from as little as one litre of bacterial culture.

We describe below protocols to express and purify several of the commonly studied members in Bcl-2 family proteins. Generally, 5–10 mg of purified protein can be obtained from 1 litre culture in approximately one week.

8.1 Human Bcl-2

Both GST- and His$_6$-tagged versions of huBcl-2 (lacking the C-terminal membrane-anchoring domain) can be produced with large yields in bacteria. For GST–Bcl-2 (1-218), we use XL-1 blue cells (Stratagene, Inc.) expressing the proteins from a pGEX vector (Pharmacia). For His$_6$–Bcl-2, we use BL21 (DE3) cells, expressing the protein from a pET vector (Novagen, Inc.).

Bcl-2 is notoriously insoluble at neutral pH. However, at pH 3.5, recombinant soluble Bcl-2 protein can be concentrated up to 10–15 mg per ml without detergents. Thus, the procedures described here are for preparation of Bcl-2 in acidic solution. The solubility of Bcl-2 at neutral pH is limited to ~50 µg per ml in PBS. High concentrations of glycerol (40%) afford little benefit at neutral pH for increasing solubility but will slow down protein aggregation. In the presence of high concentrations of non-ionic detergents (4% v/v), the concentration of Bcl-2 can reach to 1–2 mg per ml. However, the presence of detergent is not desirable for many experiments.

Protocol 22.   GST–Bcl-2 (1-218) protein

Method

11. Inoculate a colony of XL-1 blue cells carrying the pGEX–4T1–Bcl-2 (1-218) plasmids into 1 litre of Luria-Bertani (LB) medium containing 100 µg/ml ampicillin and grown at 37°C overnight with moderate agitation.

12. Dilute the culture by half in fresh LB medium containing 100 µg/ml ampicillin and allow to cool to room temperature for 1 h.

13. Induce GST–Bcl-2 protein production by addition of 0.4 M IPTG (isopropylthio-β-D-galactoside) to medium at 25°C from 6 h to over-night.

---

14. Recover cells by centrifugation using a Sorvall GSA rotor at 4000 g for 10 min. Store cells at −20°C until use. The expression level is monitored by SDS–PAGE (100 μl of the culture is resuspended in 25 μl Laemmli–SDS sample buffer, and 12.5 μl is loaded into a 15% SDS–PAGE minigel to verify protein induction: 1 μg induced band equates to ≈20 mg/l protein).

15. Resuspend cells in 50 ml of 50 mM Tris, pH 8.0, 150 mM NaCl, 1% Tween, 0.1% 2-mercaptoethonal, 5 mM EDTA, complete protease inhibitor set (Boerhinger 1697498), 1mM PMSF. Lyse cells with 0.5 mg/ml lysozyme at room temperature for 0.5 h.

16. Sonicate on ice until viscosity is minimal.

17. Centrifuge samples at 27500 g for 10 min and collect both supernatant and pellet.

18. To purify the soluble GST–Bcl-2 found in the supernatant, incubate the cell lysate with glutathionine resin (Pharmacia) at 4°C overnight. Samples should be tumbled gently during incubation in a 50 ml polypropylene centrifuge tube.

19. Pack beads in a column and wash it with 20 mM Tris, pH 8.0, 150 mM NaCl, 0.1% Tween, 0.1% 2-mercaptoethanol until the OD_{280} (optical density at 280 nm) is less than 0.01.

10. Wash beads again with 10 vols of the same buffer without detergent.

11. Elute GST–Bcl-2 with 2 vols of washing buffer containing 20 mM glutathionine. Alternatively, the beads are collected and resuspended into a 50% (v/v) slurry in washing buffer containing 20 mM glutathionine in a 50 ml polypropylene centrifuge tube, incubated at 4°C for 1 h, and the supernatant containing eluted GST–Bcl-2 is collected after centrifugation at 760 g for 5 min.

Protocol 23. His$_6$–Bcl-2 (1-218)

Method

11. Inoculate a single colony of cells into 1 litre of LB medium containing 100 μg/ml ampicillin and grow at 37°C overnight in a 2 litre Erlenmeyer flask with vigorous agitation.

12. Dilute the culture by half in fresh LB medium containing 100 μg/ml ampicillin and allow to cool to room temperature (RT) for 1 h.

13. Induce His$_6$–Bcl-2 protein production using 1 mM IPTG for 6 h at RT, with vigorous agitation.

14. Recover cells by centrifugation in a GSA rotor at 4000 g for 10 min. Store cells at −20°C until use. The expression level can be monitored by SDS–PAGE (see *Protocol 22*).

Protocol 23. *Continued*

15. Resuspend cells in 50 ml of 50 mM phosphate buffer, pH 8.0, 150 mM NaCl, 1% Tween, complete protease inhibitor set (Boerhinger 1697498), 1mM PMSF. Lyse cells with 0.5 mg/ml lysozyme at room temperature for 0.5 h.

16. Sonicate on ice until viscosity is minimal.

17. Centrifuge samples at 27 500 *g* for 10 min and collect the pellet.

18. Wash the pellet by resuspending and then centrifuging three times in 200 ml of 50 mM phosphate buffer, pH 8.0, 150 mM NaCl, 1% Tween to remove soluble contaminant proteins.

19. Resuspended pellet in 10 ml water by sonication. Add 8 M GuHCl and 1 M phosphate stock solution to give final concentrations of 6 M GuHCl, 50 mM phosphate. Adjust pH to 6.8 with NaOH.

10. Centrifuge at 27 500 *g* for 10 min and collect supernatant.

11. Add nickel resin at ~6–8 mg His_6–Bcl-2 per 1 ml resin. Add imidazole at 25 mM final concentration. Adjust the pH to 6.8.

12. Incubate at 4 °C from 3 h to overnight.

13. Pack beads in a column and wash it with 50 mM phosphate buffer, pH 6.8, 4M GuHCl, 25 mM imidazole until OD_{280} is less than 0.01.

14. Elute His_6–Bcl-2 protein with 2 × resin volume of 0.2 M acetic acid, 4 M GuHCl.

15. Dialyse the eluted His_6–Bcl-2 against 100 × volumes of cold 25 mM acetic acid, 1mM EDTA, 0.1% 2-mercaptoethonal at 4 °C for 4 h, three times.

16. Store protein at –20 °C until use.

17. The eluted sample is generally ≥90% pure. It can be further purified by gel filtration under denaturing conditions (4 M GuHCl) or, after refolding (step 15), by ion-exchange chromatography using acidic pH solution (Mono S HR10/10 at pH 4.5) (Pharmacia).

8.2 Human Bcl-X$_L$ protein

The expression and purification of human Bcl-X$_L$ protein is not difficult. Both GST–Bcl-X$_L$ and His$_6$–Bcl-X$_L$ can be expressed as soluble proteins under standard conditions. Bcl-X$_L$ is stable during thrombin digestion. Following affinity chromatography using glutathione–Sepharose or Ni-chelation resin, Bcl-X$_L$ can be further purified using Mono Q ion-exchange chromatography, typically producing a sharp peak at ≈250 mM NaCl at pH 8.0. Recombinant Bcl-X$_L$ protein is soluble in both neutral and acidic solutions. At pH 4.0, concentrations of 2–3 mg per ml of Bcl-X$_L$ protein can be obtained in the presence of 150 mM NaCl. Bcl-X$_L$ exits as monomeric protein at both neutral and acidic pH (37).

For storage of Bcl-X$_L$, samples are dialysed against 20 mM Tris, pH 8.0, 150 mM NaCl, 1mM EDTA. 0.1% 2-mercaptoethanol, and kept at 1 mg per ml at 4°C. If frozen, at least some of the Bcl-X$_L$ protein will precipitate out of solution.

Protocol 24.　GST–Bcl-X$_L$ (1-211)/Bcl-X$_L$ protein

Method

1. Inoculate a colony of XL-1 blue cells carrying the pGEX–4T1–Bcl-X$_L$ (1-211) plasmid into 1 litre of LB medium containing 100 μg/ml ampicillin and grown at 37°C overnight with moderate agitation.

2. Dilute the culture by half in fresh LB medium containing 100 μg/ml ampicillin and allow to cool to room temperature for 1 h.

3. Induce GST–Bcl-X$_L$ protein production by addition of 0.4 M IPTG to the medium at 25°C from 6 h to overnight.

4. Recover cells by centrifugation in a GSA rotor at 4000 g for 10 min. Store cells at –20°C until use. The expression level is monitored by SDS–PAGE (see *Protocol 22*) (1 μg of induced band equates to ≈20 mg/l protein).

5. Resuspend cells in 50 ml of 50 mM Tris, pH 8.0, 150 mM NaCl, 1% Tween, 0.1% 2-mercaptoethanal, 5 mM EDTA, complete protease inhibitor set (Boerhinger 1697498), 1 mM PMSF. Lyse cells with 0.5 mg/ml lysozyme at room temperature for 0.5 h.

6. Sonicate on ice until viscosity is minimal.

7. Centrifuge samples at 27 500 g for 10 min and collect the supernatant.

8. Incubate the cell lysate with glutathionine resin (Pharmacia) at 4°C overnight. The amount of resin is based on the expression level in step 4 (use ~2–4 mg recombinant protein per 1 ml beads). Samples should be tumbled gently during incubation in a 50 ml polypropylene centrifuge tube at 4°C.

9. Pack beads in a column and wash with 20 mM Tris, pH 8.0, 150 mM NaCl, 0.1% Tween, 0.1% 2-mercaptoethanol until OD$_{280}$ is less than 0.01.

10. Wash the beads again with 10 vols of the same buffer without detergent.

11. Elute GST–Bcl-X$_L$ with 2 vols of washing buffer containing 20 mM glutathionine. Alternatively, beads are collected and resuspended into a 50% (v/v) slurry in washing buffer containing 20 mM gluta-thionine in a 50 ml polypropylene centrifuge tube, incubated at 4°C

Protocol 24. *Continued*

for 1 h, and the supernatant containing eluted GST–Bcl-X$_L$ is collected after centrifugation at 760 *g* for 5 min.

12. If the GST group will be removed by proteolysis, incubate the resin after step 9 (above) with thrombin (Boerhinger) at 4°C in 20 mM Tris, pH 8.0, 150 mM NaCl, 0.1% 2-mercaptoethanol, 0.1% Tween, 2.5 mM CaCl$_2$, overnight. For 50 ml of 50% slurry, 23 mg thrombin is typically used.

13. Dialyse the eluted Bcl-X$_L$ against 100 × vols of 10 mM Tris–HCl, pH 8.0, 1 mM EDTA, 0.1% 2-mercaptoethonal at 4°C for 3 h. Inject into a Mono Q column (Pharmacia, HR10/10) that has been equilibrated with dialysis buffer, using a 2 ml/min flow rate. Wash the column with 50 ml of the same buffer at 2 ml/min. Elute Bcl-X$_L$ with a linear gradient of 0–500 mM NaCl in 60 min at 2 ml/min. Collect the samples in 1.5 ml fractions and monitor purity on 15% SDS–PAGE.

14. Pool pure fractions and store at 4°C at a concentration of 1 mg/ml.

Protocol 25. His$_6$–Bcl-X$_L$ protein

Method

11. Inoculate a single colony of cells into 1 litre of LB medium containing 100 μg/ml ampicillin and grow at 37°C overnight in a 2 litre Erlenmeyer flask with vigorous agitation.

12. Dilute the culture by half in fresh LB medium containing 100 μg/ml ampicillin and allow to cool to room temperature for 1 h.

13. Induce His$_6$–Bcl-X$_L$ protein production using 1 mM IPTG for 6 h at RT, with vigorous agitation.

14. Recover cells by centrifugation in a GSA rotor at 4000 *g* for 10 min. Store cells at –20°C until use. The expression level is monitored by SDS–PAGE (see *Protocol 22*).

15. Resuspend the cell pellet in 50 ml of 50 mM phosphate buffer, pH 8.0, 150 mM NaCl, 1% Tween, complete protease inhibitor set (Boerhinger 1697498), 1 mM PMSF. Lyse cells with 0.5 mg/ml lysozyme at room temperature for 0.5 h.

16. Sonicate on ice until viscosity is minimal.

17. Centrifuge samples at 27 500 *g* for 10 min and collect the supernatant.

18. Add nickel resin (≈6–8 mg His$_6$–Bcl-X$_L$ per 1 ml resin) in 50 mM phosphate buffer, pH 6.8, 150 mM NaCl, 1% Tween, 25 mM imidazole. Adjust the pH to 6.8. Incubate at 4°C from 3 h to overnight.

19. Pack beads into a column and wash with 50 mM phosphate buffer, pH 6.8, 150 mM NaCl, 0.1% Tween, 25 mM imidazole until OD_{280} is less than 0.01.

10. Elute His_6–Bcl-X_L protein with 2 vols of washing buffer containing 250 mM imidazole.

11. Dialyse the eluted His_6–Bcl-X_L against 100 vols of 10 mM Tris–HCl, pH 8.0, 1 mM EDTA, 0.1% 2-mercaptoethonal at 4°C for 3 h. Inject into a Mono Q column (Pharmacia, HR10/10) that was equilibrated with dialysis buffer, at 2 ml/min flow rate. Wash with 50 ml of the same buffer at 2 ml/min. Elute Bcl-X_L with a linear gradient of 0–500 mM NaCl in 60 min at 2 ml/min. Collect the samples in 1.5 ml fractions and monitor purity on 15% SDS–PAGE.

12. Pool pure fractions and store at 4°C at a concentration of ≈1 mg/ml.

8.3 Mouse Bax protein

Bax is a pro-apoptotic protein which is also toxic to *E. coli* (40). This toxicity limits the expression of Bax in *E. coli*. The protocol developed in our laboratory takes advantage of the leakiness of the *trc* promoter within pGEX plasmids to achieve a slow gradual induction, using low concentrations of IPTG and modifications of sugar levels in media. Also, the presence of a GST tag at the N-terminus of Bax can reduce the toxicity of the recombinant protein in bacteria. We have not successfully produced large amounts of His_6–Bax.

We have observed that, after cleavage and removal of GST, the recombinant Bax protein will non-specifically stick to GST resin unless ≈0.1% (v/v) non-ionic detergent is added. The concentration of purified Bax is limited to ≈0.2 mg per ml in the absence of detergent.

About half of the expressed GST–Bax protein is found in the insoluble pellet, probably associated with cell membrane debris. Non-ionic detergent does not increase the solubility of this pool of Bax. Bax protein found in the insoluble pellet, however, can be recovered through solubilization and refolding. The washed pellet is solubilized in 6 M GuHCl, 50 mM phosphate buffer, pH 7.0, 0.1% 2-mercaptoethonal. The supernatant is then diluted 30-fold into ice-cold refolding buffer (50 mM phosphate-buffer, pH 7.0, 0.1% 2-mercaptoethonal, 150 mM NaCl, 0.1% Tween-20, 1m EDTA) with vigorous stirring, then placed on ice for 16 h. After clarification by centrifugation, the diluted refolding mixture is mixed with GST–resin. This rapid dilution of GuHCl appears to be preferable to dialysis-based approaches to refolding, since when solubilized GST–Bax in GuHCl is dialysed against refolding buffer, most of GST–Bax protein appears as a high molecular weight aggregate.

A problem associated with the purification of Bax is that Bax can become cleaved during proteolytic removal of GST tag. Initially, we attempted an approach where GST–Bax was eluted and concentrated, then digested with thrombin. However, we found that most Bax is degraded under these conditions. Thus, digestion is performed directly on GST beads at 4°C for 24 h. Under these conditions, about 20% of Bax still becomes truncated near its N-terminus. If a homogeneous N-truncated Bax preparation is desirable, the eluted Bax protein from GST beads can be kept at 4°C for 1 week, which typically results in complete conversion of the Bax protein.

When chromatographed on Mono Q, Bax co-elutes at 350 mM NaCl in a broad peak. Although this ion-exchange step does not separate these proteins, it does separate Bax from contaminating thrombin and residual GST protein, and it can be used to remove detergents which have been employed in previous steps.

Purified GST–Bax, Bax, and truncated Bax all elute as high molecular weight complexes in gel filtration columns, migrating at ≈130–280 kDa. These complexes can withstand up to 4 M urea and up to 1 M NaSCN, as well as alkaline solutions to pH 9.5. In this regard, it has been reported that Bax may form heptamers (41). At low pH, Bax precipitates out of solution unless diluted into detergent-containing solutions or provided with membranes for insertion.

Protocol 26. GST–Bax (1-171) protein

Method

11. Inoculate a colony of XL-1 blue cells carrying the pGEX–4T1–Bax (1-171) plasmid into 1 litre of terrific broth medium supplemented with 1–2% glycerol containing 0.5 mg/ml cabenicillin, and grown at 37°C overnight with moderate agitation.

12. Allow the culture to cool to room temperature for 1 h, and add 10 μM IPTG.

13. Grow at room temperature for another 24 h with moderate agitation.

14. Recover cells by centrifugation in a GSA rotor at 5000 r.p.m. for 10 min. Store cells at –20°C until use. The expression level is monitored by SDS–PAGE (see *Protocol 22*).

15. Resuspend cells in 50 ml of 50 mM Tris, pH 8.0, 150 mM NaCl, 1% Tween, 0.1% 2-mercaptoethonal, 5 mM EDTA, complete protease inhibitor set (Boerhinger 1697498), 1 mM PMSF. Lyse cells with 0.5 mg/ml lysozyme at room temperature for 0.5 h.

16. Sonicate on ice until viscosity is minimal.

17. Centrifuge samples at 27 500 *g* for 10 min and collect the supernatant. Also, save the pellet if planning to attempt solubilization in 6 M GuHCl and refolding as described above.

18. Incubate cell lysate with glutathionine resin (Pharmacia) at 4°C overnight, The amount of resin is based on the expression level as determined above. Using $\approx$2–4 mg proteins per 1 ml beads. Samples should be tumbled gently during incubation in a 50 ml polypropylene centrifuge tube.

19. Pack beads in a column and wash with 20 mM Tris, pH 8.0, 150 mM NaCl, 0.1% Tween, 0.1% 2-mercaptoethanol until OD_{280} is less than 0.01.

10. Wash beads again with 10 vols of the same buffer without detergent.

11. Elute GST–Bax with 2 vols of washing buffer containing 20 mM glutathionine. Alternatively, beads can be collected and resuspended into a 50% (v/v) slurry in washing buffer containing 20 mM glutathionine in a 50 ml polypropylene centrifuge tube, incubated at 4°C for 1 h, and the supernatant containing eluted GST–Bax is collected after centrifugation at 2000 r.p.m. for 5 min.

12. If the GST group will be removed by proteolysis, incubate the resin after step 9 (above) with thrombin (Boerhinger) at 4°C in 20 mM Tris, pH 8.0, 150 mM NaCl, 0.1% 2-mercaptoethanol, 0.1% Tween, 2.5 mM $CaCl_2$ overnight. For 50 ml of 50% slurry, 23 mg thrombin is used. Elute Bax with 2 $\times$ resin volumes of the same buffer.

13. Dialyse the eluted Bax protein against 100 vols of 10 mM Tris–HCl, pH 8.0, 1 mM EDTA, 0.1% 2-mercaptoethonal at 4°C for 3 h. Inject into a Mono Q column (Pharmacia, HR10/10) that has been equilibrated with dialysis buffer, using a 2 ml/min flow rate. Wash the column with 50 ml of same buffer at 2 ml/min. Elute Bax with a linear gradient of 0–500 mM NaCl in 60 min at 2 ml/min. Collect the samples in 1.5 ml fractures and monitor purity on 15% SDS–PAGE.

14. Pool pure fractions and store at –20°C at a concentration of $\approx$0.2 mg/ml.

8.4 Human Bid protein

The expression of Bid in bacteria is easily achieved. The recombinant BID protein is highly soluble at both neutral and acidic pH (37). The only problem associated with purification of Bid is that it is susceptible to digestion by thrombin. Therefore, performing partial digestions of GST–Bid is advised to reduce the degradation of Bid by thrombin. We typically digest GST–Bid under the same conditions as those described above for other Bcl-2 family proteins, except the digestion is stopped after 1–2 hours by addition of PMSF. The eluted protein is then dialysed against 50 mM acetate buffer, pH 4.8, 1 mM EDTA, 0.1% 2-mercaptoethonal, and Mono S ion-exchange chromatography is performed. Bid elutes at very low salt concentrations ($\sim$50 mM

NaCl), whereas the degraded contaminant elutes at high concentrations of salt. The recombinant purified Bid protein exists as monomeric protein at neutral and acidic pH (37).

Protocol 27.   GST–Bid/Bid protein

Method

11. Inoculate a colony of XL-1 blue cells carrying the pGEX–4T1–Bid plasmid into 1 litre of terrific broth medium supplemented with 1–2% glycerol and containing 0.5 mg/ml cabenicillin. Grown at 37 °C overnight with moderate agitation.

12. Allow the culture to cool to room temperature for 1 h. Add 0.4 mM IPTG and grow at room temperature for 6 h with moderate agitation.

13. Recover cells by centrifugation in a GSA rotor at 4000 g for 10 min. Store cells at –20 °C until use. The expression level is monitored by SDS–PAGE (see *Protocol 22*).

14. Resuspend the cell pellet in 50 ml of 50 mM Tris, pH 8.0, 150 mM NaCl, 1% Tween, 0.1% 2-mercaptoethonal, 5 mM EDTA, complete protease inhibitor set (Boerhinger 1697498), 1 mM PMSF. Lyse cells with 0.5 mg/ml lysozyme at room temperature for 0.5 h.

15. Sonicate on ice until viscosity is minimal.

16. Centrifuge samples at 27 500 g for 10 min. Collect the supernatant and discard the pellet.

17. Incubate the cell lysate with glutathionine resin (Pharmacia) at 4 °C overnight. The amount of resin is based on the expression level. Use ≈2–4 mg protein per 1 ml beads). Samples should be tumbled gently during incubation in a 50 ml polypropylene centrifuge tube.

18. Pack beads in a column and wash with 20 mM Tris, pH 8.0, 150 mM NaCl, 0.1% Tween, 0.1% 2-mercaptoethanol until OD_{280} is less than 0.01.

19. Wash beads again with 10 vols of the same buffer without detergent.

10. Elute GST–Bid with 2 × resin volumes of washing buffer containing 20 mM glutathionine. Alternatively, beads are collected and resuspended into a 50% (v/v) slurry in washing buffer containing 20 mM glutathionine in a 50 ml polypropylene centrifuge tube, incubated at 4 °C for 1 h, and the supernatant containing eluted GST–Bid is collected after centrifugation at 760 g for 5 min.

11. If the GST group will be removed by proteolysis, incubate the resin after step 9 (above) with thrombin (Boerhinger) at 4 °C in 20 mM Tris, pH 8.0, 150 mM NaCl, 0.1% 2-mercaptoethanol, 0.1% Tween, 2.5 mM

CaCl$_2$ for 2 h, and stop the reaction by adding 1 mM PMSF and 5 mM EDTA. For 50 ml of 50% slurry, 23 mg thrombin is used. Elute Bid with 2 × resin volumes of the same buffer.

12. Dialyse the eluted Bid against 100 vols of 50 mM acetate buffer, pH 4.8, 1mM EDTA, 0.1% 2-mercaptoethonal at 4°C overnight. Inject into a Mono S column (Pharmacia, HR10/10) that has been equilibrated with dialysis buffer using a 2 ml/min flow rate. Wash with 50 ml of the same buffer at 2 ml/min. Elute Bid with a linear gradient of 0–500 mM NaCl in 60 min at 2 ml/min. Collect the samples in 1.5 ml and monitor purity on 15% SDS–PAGE.

13. Pool pure fractions and store at –20°C at a concentration of ≈1 mg/ml.

9. Measurements of pore formation by Bcl-2 family proteins

One of the biochemical activities ascribed to some members of the Bcl-2 family of proteins is ion-channel activity (reviewed in ref. 42). The determination of the three-dimensional structure of Bcl-X$_L$ (43) showed that this protein shares striking similarities with the previously determined structures of the pore-forming domains of the bacterial toxins diphtheria (44) and colicins A, E1, and Ia (45–47). These bacterial proteins share the same unique ability in that they can exist either as soluble or membrane-inserted proteins. They are able to achieve this large change in character by burying two, long, predominantly hydrophobic α-helices within a shell of amphipathic helices. Under appropriate conditions, the outer amphipathic α helices splay away, freeing the central hydrophobic α helical hairpin to insert into the membrane bilayer which initiates channel formation.

Bcl-X$_L$ consists of a bundle of seven α helices, arranged in three layers. The outer two layers of amphipathic helices shield between them two long central α-helices, each ~20 amino acids in length with a bias towards hydrophobic residues (43). The two central α helices are of sufficient length to span the hydrophobic cross-section of a membrane bilayer. The structural similarity between Bcl-X$_L$ and the pore-forming domains of the bacterial toxins suggests that the Bcl-2 protein family may possess pore-forming potential as well. Indeed, several studies have shown that Bcl-2, Bcl-X$_L$, and Bax exhibit channel-forming activity *in vitro* (36, 48–50).

The following section describes methods for examining Bcl-2 protein family channel formation on a macroscopic level. Macroscopic level measurements have the advantage in that in order for channel activity to be detected, a majority of the protein molecules must participate in channel formation. However, the methods described do not yield information about channel formation on a single-channel basis, particularly in terms of conductivity and ion selectivity. For quantifying these characteristics, planar bilayer measurements are more appropriate. Explanation of this technique may be found in several reviews, including refs 51 and 52.

9.1 Liposome preparation

9.1.1 Lipids

Two lipid sources are Avanti Polar Lipids (Birmingham, AL) and Sigma (St. Louis, MO). The lipids are supplied either in powder or chloroform suspensions. In either case, the contact with air of these lipid stocks should be kept to a minimum to prevent lipid oxidation.

The lipids may be either synthetic or purified from a natural source, such as the soybean lipid, asolectin. The latter is a mosaic of lipids, and in the case of colicin E1, higher activity is often observed over liposomes produced from synthetic lipids (53). The synthetic lipids allow greater control over vesicle composition, particularly the ratio of charged to uncharged lipids.

Studies of both bacterial toxins and the Bcl-2 protein family have utilized lipid vesicles that contain at least 10 mol% negatively charged lipids. The negatively charged lipid most commonly used has a glycerol moiety at its headgroup, although phosphatidylserine carries a negative charge as well. In the case of colicin E1, the net negative charge of the vesicle is important for the tight binding of the protein to the membrane surface (54). Presumably, this requirement may be applied to the Bcl-2 protein family as well.

9.1.2 Lipid purification

Synthetic lipid stocks are of sufficient purity to be used as supplied. Further purification of asolectin is often performed. The following purification procedure is a modification of that described previously (55).

Protocol 28.

Method

1. 10 g asolectin is added to 100 ml acetone and stirred overnight.

2. The mixture is centrifuged at 5000 *g* for 10 min at room temperature.

3. The supernatant is discarded, the pellet dissolved in 40 ml anhydrous ether, and the mixture centrifuged for 10 min at 5000 *g*, at room temperature.

4. The ether is evaporated on a rotary evaporator device.

5. The dried lipid is dissolved in ~80 ml chloroform.

6. The lipid concentration is determined by a phosphate analysis (56).

9.1.3 Liposome production

For solute efflux to be measured, the liposomes must be large (> 0.1 mm diameter) and unilamellar, i.e. having only a single membrane bilayer. To produce such vesicles, the following procedure, a modification of that outlined in Peterson and Cramer (57) is employed.

Protocol 29.

Method

1. The appropriate amount of lipid is measured. Typically, a liposome suspension having the final concentration of 10 mg/ml lipid is used, of which ~80 and 20% are neutral and negatively charged lipids, respectively. This ratio can be varied. If powdered lipid stocks are used, the lipid powder is first dissolved in chloroform. The chloroform lipid suspension should be placed in a Pyrex tube.

2. The chloroform is evaporated under a stream of either nitrogen or argon while vortexing. Use of a vortex ensures an even distribution of the lipid film. Be certain that no pockets of chloroform are entrapped under a thin layer of dried lipid, as inadequate removal of organic solvents will affect liposome formation. The lipid film is further dried under vacuum for at least two hours to remove any remaining trace of organic solvent.

3. The lipid is then resuspended in buffer to the desired final concentration (typically 10 mg/ml). The buffer utilized in this laboratory is 10 mM DMG (dimethyl glutaric acid), 100 mM KCl, 2 mM $Ca(NO_3)_2$, pH 5.0, although Hepes or phosphate buffers can be employed. The suspension is vortexed under a stream of N_2 for at least 5 min to flush out any air from the tube. The tube is then sealed with at least two layers of Parafilm, the top layer being somewhat lose.

4. The tube is placed in a sonicator bath (Branson). For best results, the bath should be cleaned well and allowed to run for 30 min prior to lipid sonication. The tube should be placed in a visible 'node' with the liquid meniscus even with the water level. Continue sonication until the solution has been converted from opaque to completely clear. The lipid suspension is now composed of small unilamellar vesicles (SUVs).

5. To convert the SUVs to large unilamellar vesicles (LUV) several freeze–thaw cycles are employed. The suspension is frozen as rapidly as possible in a dry-ice ethanol bath or liquid nitrogen. The frozen solution is allowed to thaw slowly to room temperature. Repeat with at least four additional freeze–thaw cycles. The solution should again be opaque.

9.2 Channel activity measurements

9.2.1 Electrode set-up

Changes in chloride concentration are followed with a chloride-specific electrode (Orion 94-17B) coupled to a double junction reference electrode (Orion 90-02). The electrode signal is output to a pH meter or other suitable voltmeter attached to a strip chart recorder. The meter utilized in this laboratory was built according to the method of Cramer and colleagues (57) which allows for expansion of the scale and readings of very small concentration changes. In addition to chloride efflux, channels formed by the Bcl-2 protein family will also permit passage of K^+, and a K^+-sensitive electrode (Orion 93-19) may be easily substituted for the chloride electrode.

9.2.2 Extravesicular buffer and protein buffers

The buffer commonly used for chloride efflux measurements in this laboratory is that described in Peterson and Cramer (57). The buffer composition is 10 mM DMG, 100 mM choline nitrate, 2 mM $Ca(NO_3)_2$. Choline nitrate is formed from choline bicarbonate (Sigma) and nitric acid. The buffer is degassed for at least 1 h and titrated to the appropriate pH with NaOH. Other chloride-free buffers are also suitable for measurement, provided their ionic strength is similar to that used for vesicle preparation.

The protein sample should be in a buffer that is chloride free, if possible, and should also lack any compounds which would interfere with electrode response. β-Mercaptoethanol has been found to especially perturb electrode response (S. L. Schendel and J. C. Reed, unpublished results).

9.2.3 Ion efflux measurement

Extravesicular buffer (15 ml) is placed in a disposable plastic beaker (Fisher). Use of such inexpensive beakers removes the need for tedious washing between trials. Approximately 75–100 ml of the liposome suspension is pipetted into the beaker and the electrode allowed to equilibrate until a stable baseline is achieved. The solution should be stirred at medium speed at all times and no bubbles should adhere to the electrode tips. Once the baseline has stabilized, 2 ml of a 70 mM methanol stock of the K^+-specific ionophore valinomycin is added to the beaker. Valinomycin induces the formation of a Nernst diffusion potential of ~135 mV, negative inside. The ionophore may be omitted if no membrane potential is desired. The electrode is again allowed to equilibrate following valinomycin addition. The Bcl-2 family protein is then added to a concentration that will induce release of 50–80% of the total encapsulated Cl^- and induce a response having a slope between 30 and 60° which facilitates accurate calculation of the efflux rate. After the protein-induced efflux has levelled off, Triton X-100 is added to the beaker to a final concentration of 0.1%. This detergent addition lyses the vesicles and releases any residual chloride, allowing the total amount of chloride encapsulated to be determined. Examples of chloride-efflux induced by Bcl-2 family proteins, as measured by these methods, can be found in previous publications (36, 38, 48).

At the end of the measurements, a calibration curve is produced by additions of known amounts of KCl. The KCl is added in the presence of an aliquot of liposomes, so that the background Cl⁻ concentration is similar. KCl is added in progressively increasing amounts until the chart recorder deflection equals the maximum observed by Triton X-100 lysis of the vesicles.

For colicin molecules, where the molecularity of the channel is known to be 1 (58), a specific Cl⁻ efflux rate can be calculated by determining the ratio of colicin-induced Cl⁻ release to the total encapsulated Cl⁻. This value is then divided by the number of colicin molecules present (assuming that each molecule forms one channel) and the quotient divided by the time taken for the efflux to occur. Since the molecularity of the Bcl-2 protein family channels is still unclear, this rate calculation is not as straightforward. For purposes of comparison between trials, this laboratory simply determines the slope of the initial response.

10. Methods of assaying Bcl-2 and Bax family protein function in yeast

Bcl-2 family proteins play an evolutionarily conserved role in regulating the life and death of the cell. Certain pro-apoptotic members of the Bcl-2 family, Bax and Bak, have intrinsic cytotoxic activities in that they not only induce or sensitize mammalian cells to undergo apoptosis but also display a lethal phenotype when ectopically expressed in two yeast species *Saccharomyces cerevisiae* and *S. pombe* (59–65). Furthermore, the anti-apoptotic Bcl-2 and Bcl-X$_L$ proteins can protect yeast against Bax-mediated lethality, suggesting that the death-regulatory functions of these Bcl-2 family proteins are well preserved in yeast. These observations provide the opportunity to study the function of Bcl-2 family proteins in genetically tractable yeast and to apply classical yeast genetics and functional cloning approaches to the dissection of programmed cell death pathway regulated by Bcl-2 family proteins. We describe here methods used in our laboratory to assess cytotoxic and cytoprotective functions of Bcl-2 family proteins in the budding yeast *S. cerevisiae*.

10.1 Plasmid considerations

A variety of expression vectors are available to express foreign genes in *S. cerevisiae*. Two major issues require consideration when choosing a yeast vector for expressing Bax or other Bcl-2 family members in these organisms. First, Bax needs to accumulate to high levels to cause cell death, thus a strong promoter should be employed. Secondly, vectors with appropriate selectable markers should be chosen depending on the yeast strain used. Two general types of promoters are used to achieve high level gene expression in *S.*

cerevisiae: (a) constitutive and (b) conditional promoters. Promoters of the alcohol dehydrogenase (*ADH1*) and glyceraldehyde-3-phosphate dehydrogenase (*GPD* or *GAP*) genes are commonly employed strong constitutive promoters. When expressing the *bax* gene under the control of either the *ADH1* or *GPD* promoter, the diminished number of transformed colonies can be employed as an indicator of the lethality caused by Bax. A parallel transformation with an 'empty' vector is included as a control. This assay, however, has limitations in interpreting the data, because reductions in transforming activity can be a result of cell cycle arrest, cell death, or even some contaminants in DNA that affect transformation efficiency. For this reason, we prefer a regulatable expression system over a constitutive system. Several conditional promoters have been employed for expressing heterologous genes in yeast; the most commonly used is a hybrid promoter combining the *UAS* (upstream activating sequence) from the *GAL1/GAL10* gene and the basal promoter from the *CYC1* gene, these are often referred to as the *GAL1* or *GAL10* promoters. The *GAL* promoter is regulated by the availability of carbon sources in the growth medium: transcription is repressed when yeast are grown in glucose-containing medium; derepressed when yeast are grown in raffinose-containing medium; and dramatically induced when yeast are grown in galactose-containing medium. Under optimal conditions, inductions as high as a 1000-fold can be achieved when yeast are grown in medium which contains galactose but lacks glucose. The optimal time for inducing transcription from the *GAL* promoter is strain dependent and thus should be determined empirically. In general, however, we perform assays at 18–24 h after placing cells in galactose-containing medium. The induction can be monitored by immunoblot analysis using antibodies directed against the heterologous protein of interest.

Basic machinery for transcription and translation is highly conserved from yeast to mammals; therefore, a mammalian cDNA, with minor engineering, often expresses well in yeast. By employing a yeast promoter, transcription initiation of the heterologous gene should proceed efficiently. Most expression vectors also include a yeast transcription terminator, such as the terminator for the *CYC1* gene, to ensure production of correctly sized mRNAs and to improve transcription efficiency. Translation initiation in mammalian cells occurs most efficiently within the context of preferred sequences flanking the AUG initiation codon, commonly referred to as a Kozak sequence (66). AUG codons that deviate from the Kozak consensus often translate inefficiently. No well-defined equivalent of the Kozak sequence exists in yeast. To achieve efficient translation initiation, it is required that the initiation codon within heterologous cDNAs represents the first AUG from the 5'-end. In addition, it is preferable to eliminate as much of the 5'-untranslated sequences from the introduced cDNA as possible, since translation can be hindered if the 5'-untranslated region of the mRNA has the potential to form secondary structures or contains a high GC content (67).

For our experiments, we employed the *GAL* promoter-containing vector YEp51 for expressing Bax in *S. cerevisiae*. YEp51 is a high copy number yeast plasmid bearing the *LEU2* gene as a selectable marker. We cloned *bax* cDNA into YEp51 using the SalI site in the polylinker (i.e. the restriction site closest to the *GAL10* promoter), thus keeping the length of the resulting 5'-untranslated region to a minimum (61).

A variety of yeast expression vectors containing either the *GAL* regulatable promoter or the *ADH1* and *GPD* constitutive promoters are available from American Type Culture Collection (ATCC, Manassas, VA) (68, 69).

10.2 *Saccharomyces cerevisiae* strains

Once constructed, the expression plasmid is transformed into an appropriate yeast strain following standard transformation procedures, as described below. In principle, any laboratory strain with appropriate auxotrophic mutations corresponding to the selectable marker on the expression plasmid can be used as a host strain for expressing Bax. In practice, however, strain background variations may affect the sensitivity of yeast to Bax. We have tested the effect of Bax in a few haploid strains. Some strains, such as BF264-15Dau (70), were found to be relatively more sensitive to Bax, suggesting that strain differences do matter. Thus, it is advisable to test yeast stains from different origins, so that the most sensitive ones can be employed. A second factor to consider when selecting a yeast strain is the availability of auxotrophic markers. *URA3*, *LEU2*, *HIS3*, and *TRP1* are the four commonly used auxotrophic markers for *S. cerevisiae*. Strains with multiple auxotrophic markers provide greater flexibility, especially in cases where the effects of more than one protein are to be tested. Some commonly used laboratory yeast strains contain these auxotrophic mutations, including W303a or a (genotype: *MAT* **a**/or *a leu2–3 112 his3–11,15 try1–1 ura3–1 can1–100 ade2–1*) and EGY48 (genotype: *MAT***a** *leu2–3 112 his3–11,15 trp1–1 ura3–1 6LexAop-LEU2*). Yeast strains can be obtained from the Yeast Genetic Stock Center at ATCC.

10.3 Media and growth conditions

We routinely culture yeast in YPD-rich medium. Once a plasmid is introduced, yeast should be grown in synthetic drop-out medium to select for the plasmid. For example, after transforming YEp51–Bax into yeast, transformants are initially selected and subsequently maintained on synthetic drop-out medium lacking leucine (SD–Leu) and supplemented with glucose to repress Bax expression. We observed that the amount of nutrients in the medium can also affect the sensitivity of yeast to pro-apoptotic proteins. In general, we find that yeast are more sensitive to Bax when cultured in medium with minimal nutrients. For this reason, Burkholder's Minimal Medium (BMM) medium (71) may be useful for detecting a more striking lethal effect of Bax in yeast. The preparation of commonly used media is listed below (see refs 72 and 73 for detailed descriptions of *S. cerevisiae* media).

Protocol 30. YPD medium

Method

1. To a 2 litre flask, add 10 g Bacto-yeast extract and 20 g Bacto-peptone. Include 20 g of agar when making YPD plates.

2. Bring the volume to 900 ml with distilled H_2O.

3. Following autoclaving, add 100 ml of filter-sterilized 20% glucose (dextrose) stock.

4. When making plates, let the medium cool to 65°C and then pour into Petri dishes. 1 litre of medium should yield about 50 100 mm diameter plates.

Protocol 31. SD medium

There are three steps to the preparation of SD medium.

Method

1. Prepare the following amino acid and supplement stocks in distilled H_2O: 4 mg/ml each of adenine, L-histidine–HCl, L-arginine–HCl, L-methionine, L-leucine, L-isoleucine, L-lysine–HCl, phenylalanine, aspartic acid, and valine; 2 mg/ml of uracil. Sterilize by autoclaving. 4 mg/ml each of L-tryptophan, L-tyrosine, and L-threonine and sterilize by filtration. Store at room temperature.

2. To a 2-litre Erlenmeyer flask, add 6.7 g yeast nitrogen base without amino acids, and 10 ml each of the above amino acid stocks and 20 ml of uracil, leaving out the three heat-sensitive amino acids (tryptophan, tyrosine, and threonine). Omit the amino acid(s) that will be used for selection of the plasmid(s). Add 20 g of agar when making solid medium. Bring the volume to 870 ml with distilled H_2O and autoclave.

3. Cool the solution to about 80°C, then add 10 ml each of the three heat-sensitive amino acids and 100 ml of filter-sterilized sugar stocks (either 20% glucose or 20% galactose; we often supplement galactose medium with 1% raffinose). SD plates are prepared similarly as described for YPD plates above.

Protocol 32.   BMM medium

Method

1. Prepare the following stock solutions in distilled H_2O.
Mineral stocks (10 000 ×)

   (a) To a 100-ml Erlenmeyer flask, add 30 mg H_3BO_3, 50 mg $MnSO_4.7H_2O$, 150 mg $ZnSO_4.7H_2O$, and 20 mg $CaSO_4.5 H_2O$. Bring the volume to 50 ml with dH_2O.

   (b) To a 100-ml Erlenmeyer flask, add 100 mg $Na_2MoO_4.2H_2O$. Bring the volume to 50 ml with dH_2O.

   (c) To a 100-ml Erlenmeyer flask, add 125 mg $FeCl_3.6H_2O$. Bring the volume to 50 ml with dH_2O.

   (d) Do not autoclave. Store at 4°C.

Vitamin stock (1000 ×)
To a 250-ml Erlenmeyer flask, add 20 mg thiamine, 20 mg pyridoxine, 20 mg nicotinic acid, 20 mg pantothenic acid, 0.2 mg biotin, and 1 g inositol, add H_2O to 100 ml. Filter sterilize and store at 4°C.

2. To a 2-litre Erlenmeyer flask, add 1.5 g KH_2PO_4, 500 mg $MgSO_4.7H_2O$, 330 mg $CaCl_2.2H_2O$, 100 µg KI, 2 g asparagine. 100 µl each of mineral stocks (a), (b), and (c) (10000 ×) and 10 ml each required amino acids (the ones that yeast can not make) except the heat-sensitive ones (amino acid stocks are made as described for SD medium). Include 20 g agar when making solid medium. Bring the volume to 900 ml with dH_2O and autoclave.

3. Following autoclaving and cooling to about 80°C, add 1 ml vitamin stock (1000 ×), 10 ml each of heat-sensitive amino acids and 100 ml of appropriate carbon source (from 20% glucose or galactose stock as described for SD medium). BMM plates are made similarly as described for YPD or SD plates.

10.4 Transformation of *S. cerevisiae* by the LiOAc method

Plasmid DNA can be introduced into yeast by several different means. We routinely transform *S. cerevisiae* using the LiOAc method (74). The transformation efficiency achieved with this method can be influenced by multiple factors. Strain background, cell density at the time of transformation, the quality of the plasmid DNA and carrier DNA used for transformations, and the length of heat-shock can all play some role in influencing the final outcome. When using the protocol below, the transformation efficiency we routinely obtain is 10^3–10^4 transformants/ µg DNA.

Protocol 33.

Method

1. Inoculate a single colony into 50 ml YPD or SD medium in a 250-ml Erlenmeyer flask and grow yeast to a mid-exponential phase ($OD_{600}=0.8$) at 30°C in an incubator-shaker set at 250 rpm.

2. Harvest cells by centrifugation (5 min at 1000 g) and wash once in 50 ml distilled H_2O.

3. Collect cells by centrifugation as above. Wash cells in 10 ml LiOAc solution (0.1 M LiOAc, 10 mM Tris–HCl, pH 8.0, 1 mM Na_2EDTA) and resuspend the cell pellet in 0.5 ml (1/100 of initial culture volume) LiOAc solution. Cells can be stored in LiOAc at 4°C for up to one week and transformation can be performed at a later time, but with reduced efficiency.

4. Aliquot 100 μl cell suspension into each microfuge tube and add 5 μl 10 mg/ml carrier DNA (sheared single-strand salmon sperm DNA; Sigma) and 1 mg plasmid DNA to each aliquot. Keep the total volume of DNA below 10 ml.

5. Add 0.7 ml PEG solution [40% polyethylene glycol (PEG 3300–4000), 0.1 M LiOAc, 10 mM Tris–HCl, pH 8.0, 1 mM Na_2EDTA, pH 8.0] to the cell DNA mixture, vortex vigorously, and incubate at 30°C for 45 min with constant shaking.

6. Heat-shock in a 42°C water bath for 5–15 min. Centrifuge in a microcentrifuge at 1000 g for 5 min and resuspend the cells in 200 μl TE buffer (10 mM Tris–HCl, pH 8.0, 1 mM Na_2EDTA, pH 8.0).

7. Spread the transformation mixture on SD solid plates lacking amino acids required for selection of the plasmid and place the plates in a 30°C incubator (VWR Scientific Products).

8. Colonies should appear in 2–4 days.

10.5 Assay for Bax-induced lethality in *S. cerevisiae*

When YEp51–Bax is introduced into an appropriate yeast strain, the effect of Bax on yeast growth can be easily observed by plate tests. Yeast containing YEp51–Bax are maintained on glucose-containing SD–Leu plates. A single colony can be streaked on galactose-containing SD–Leu plates to induce Bax expression side by side with a control streak from a colony containing YEp51 lacking Bax ('empty'). This test of growth inhibition can also be performed by standard replica-plating, although it is less sensitive than streaking. The drastically reduced growth rate observed on galactose medium which induces the *bax* gene indicates that the gene product is either cytotoxic to yeast or induces cell cycle arrest. Two assays can be employed to distinguish between these possibilities: the clonogenic (or colony formation) assay and the vital dye exclusion assay.

10.5.1 Clonogenic assay

The rationale for this assay is that if inducing Bax expression (by culturing in galactose medium) induces cell death there will be few colonies formed when spreading these dead cells on glucose plates since death is irreversible. On the other hand, if Bax simply causes cell cycle arrest (a reversible process), cells should be able to re-enter the cell cycle and to proliferate when shifted back to permissive conditions (glucose medium) where Bax expression is shut off.

1. Inoculate a single colony of yeast bearing YEp51–Bax into 5 ml liquid SD–Leu medium and culture at 30°C in an incubator-shaker (Lab-line Instruments Inc., Melrose Park, IL) set at 250 r.p.m. to a mid-exponential phase (OD_{600}=0.4–0.6).

2. Cells are harvested by centrifugation at 1000 g for 5 min and washed three times in sterilized H_2O or PBS at room temperature to remove residual glucose.

3. After the final wash, cells are resuspended in 5 ml galactose-containing SD–Leu liquid medium for continuous culturing. At different times thereafter, cell densities are determined by measuring OD_{600}, (OD_{600} of 1 equals about 3×10^7 cells/ml). Cells are then diluted and about 10^3 cells are spread on glucose-containing SD–Leu plates.

10.5.2 Vital dye exclusion assay

Cell viability can also be determined by the vital dye exclusion assay. Dead cells have lost plasma membrane integrity and as a result cannot exclude dyes and thus are stained. For example, trypan blue can be used to stain yeast. The ability of cells to exclude trypan blue indicates that they are still alive. Thus, if yeast failed to grow on galactose plates but still excludes trypan blue, then presumably this is the result of a problem with the cell cycle rather than cell death. On the other hand, inability to exclude trypan blue (cells are thus stained blue) indicates that the gene introduced is lethal to yeast. The percentage of blue cells can also be employed as an indicator of the potency of the killing. Other vital dyes, such as methylene blue and erythrosine B, have also been employed (62).

1. Grow YEp51–Bax-containing yeast and transfer to galactose-containing medium as described above for the clonogenic assay.

2. At different times after inducing Bax expression in galactose-containing liquid medium, remove 20 μl of the yeast culture and mix with 20 μl of 0.4 % of trypan blue solution in PBS (Sigma). The numbers of dead (blue) and live (no colour) cells are determined using a haemocytometer (Hausser

Scientific, Horsham, PA) under a light microscope and the percentage of dead cells is calculated. The existence of cell wall on yeast does not interfere with trypan blue assays.

When the death gene introduced is under the control of a constitutive promoter, e.g. the *ADH1* or *GPD* promoter, no transformants or far fewer transformants will form after introducing the expression plasmids into yeast and plating on selective drop-out medium, which selects for the introduced plasmid. An 'empty' vector should also be transformed in parallel as a control. In this assay, however, one will not be able to tell the difference between growth arrest and cell death. One way to circumvent this, is to introduce, for example, *ADH1–Bax*-bearing plasmid into a strain already containing an anti-apoptotic gene under the control of a conditional promoter, e.g. *GAL–bcl-2*. When cells are cultured in galactose-containing medium, expression of the anti-apoptotic gene will protect yeast cells from the cytotoxic effect of the introduced *bax* gene. After transformants are formed on galactose plates, these can be streaked or patched on glucose-containing plates where expression of the anti-apoptotic *bcl-2* gene is shut off, and thus the effect of the *bax* gene is revealed. In this case, streaking or patching in parallel to a glucose plate provides a control for any non-specific growth failures due to unforeseen technical problems.

11. Summary

The methods described here provide a broad range of techniques for scientists interested in studying the expression and function of Bcl-2 family proteins. Though alternative approaches are possible, the protocols presented here have proved robust and reproducible when used by multiple investigators in our laboratory. We welcome feedback from others who attempt these methods or who make further modifications and improvements in the future.

Acknowledgements

We thank T. Brown for manuscript preparation and E. Smith for artwork.

References

1. Reed, J. (1998) *Oncogene* **17,** 3225–3236
2. Zamzami, N., Brenner, C., Marzo, Susin, S., and Kroemer, G. (1998) *Oncogene* **16,** 2265–2282
3. Chao, D.T., and Korsmeyer, S. J. (1998) *Annu. Rev. Immunol.* **16,** 395–419
4. Adams, J., and Cory S. (1998) *SCIENCE* **281,** 1322–1326
5. Reed, J. C. (1996) *Behring Inst. Mitt.* **97,** 72–100
6. Reed, J. C. (1997) in *Vitamins and Hormones* (Litwack, G., ed) Vol. 53, pp. 99–138, Academic Press, San Diego
7. Reed, J. (1998) *Heart Failure Rev.* **3,** 15–26
8. Thompson, C. B. (1995) *SCIENCE* **267,** 1456–1462
9. Strasser, A., Huang, D. C. S., and Vaux, D. L. (1997) *Biochim. Biophys. Acta* **1333,** F151–F178

10. Krajewski, S., Zapata, J. M., and Reed, J. C. (1996) *Anal. Biochem.* **236,** 221–228
11. Smith, P., Krohn, R., Hermanson, G., Mallia, K., Gartner, F., Prozenzano, M., Fujimoto, E., Goeke, N., Olson, B., and Klenk, D. (1985) *Anal. Biochem.* **150,** 76–85
12. Krajewski, S., Bodrug, S., Gascoyne, R., Berean, K., Krajewska, M., and Reed, J. C. (1994) *Am. J. Pathol.* **145,** 515–525
13. Krajewski, S., Krajewska, M., Shabaik, A., Wang, H.-G., Irie, S., Fong, L., and Reed, J. C. (1994) *Cancer Res.* **54,** 5501–5507
14. Krajewski, S., Krajewska, M., Shabaik, A., Miyashita, T., Wang, H.-G., and Reed, J. C. (1994) *Am. J. Pathol.* **145,** 1323–1333
15. Krajewski, S., Bodrug, S., Krajewska, M., Shabaik, A., Gascoyne, R., Berean, K., and Reed, J. C. (1995) *Am. J. Pathol.* **146,** 1309–1319
16. Krajewski, S., Blomvqvist, C., Franssila, K., Krajewska, M., Wasenius, V.-M., Niskanen, E., and Reed, J. C. (1995) *Cancer Res.* **55,** 4471–4478
17. Krajewski, S., Mai, J. K., Krajewska, M., Sikorska, M., Mossakowski, M. J., and Reed, J. C. (1995) *J. Neurosci.* **15,** 6364–6376
18. Krajewski, S., Krajewska, M., and Reed, J. C. (1996) *Cancer Res.* **56,** 2849–2855
19. Kitada, S., Krajewska, M., Zhang, X., Scudiero, D., Zapata, J. M., Wang, H. G., Shabaik, A., Tudor, G., Krajewski, S., Myers, T. G., Johnson, G. S., Sausville, E. A., and Reed, J. C. (1998) *Am. J. Pathol.* **152**(1), 51–61
20. Pezzella, F., Tse, A., Cordell, J., Pulford, K., Gatter, K., and Mason, D. (1990) *Am. J. Pathol.* **137,** 225
21. Reed, J. C., Tanaka, S., Cuddy, M., Cho, D., Smith, J., Kallen, R., Saragovi, H. U., and Torigoe, T. (1992) *Anal. Biochem.* **205**(1), 70–76
22. Hockenbery, D. M., Zutter, M., Hickey, W., Nahm, M., and Korsmeyer, S. J. (1991) *Proc. Natl. Acad. Sci. USA* **88,** 6961–6965
23. Novack, D. V., and Korsmeyer, S. J. (1994) *Am. J. Pathol.* **145,** 61–73
24. Broome, H. E., Dargan, C. M., Bessent, E. F., Krajewski, S., and Reed, J. C. (1995) *Immunol.* **84,** 375–382
25. Tsujimoto, Y., and Croce, C. (1986) *Proc. Natl .Acad. Sci. USA* **83,** 5214–5218
26. Negrini, M., Silini, E., Kozak, C., Tsujimoto, Y., and Croce, C. M. (1987) *Cell* **49,** 455–463
27. Chomczynski, P., and Sacchi, N. (1987) *Anal. Biochem.* **162,** 156–159
28. Chomczynski, P. (1992) *Anal. Biochem.* **201,** 134–139
29. Ogretmen, B., Ratajczak, H., Kats, A., Stark, B. C., and Gendel, S. M. (1993) *Biotechniques* **14,** 932–935
30. Maniatis, T., Fritsch, E. F., and Sambrook, J. (1982) *Molecular Cloning*, 0 Ed., Cold Spring Harbor Laboratory, Cold Spring Harbor
31. Kitada, S., Miyashita, T., Tanaka, S., and Reed, J. C. (1993) *Antisense Res. Dev.* **3,** 157–169
32. Kitada, S., Takayama, S., DeRiel, K., Tanaka, S., and Reed, J. C. (1994) *Antisense Res. Dev.* **4,** 71–79
33. Miyashita, T., Krajewski, S., Krajewska, M., Wang, H. G., Lin, H. K., Hoffman, B., Lieberman, D., and Reed, J. C. (1994) *Oncogene* **9,** 1799–1805
34. Wijsman, J.H., Jonker, R.R., Keijzer, R., van de Velde, C.J.H., Cornelisse, C.J., and van Dierendock, J.H. (1993) *J. Histochem. Cytochem.* **41,** 7–12
35. Sternberger, L. A., Hardy Jr., P. H., Cuculis, J. J., and Meyer, H. G. (1970) *J. Histochem. Cytochem.* **18,** 315–320
36. Schendel, S. L., Xie, Z., Montal, M. O., Matsuyama, S., Montal, M., and Reed, J. C. (1997) *Proc. Natl. Acad. Sci. USA* **94,** 5113–5118
37. Xie, Z., Schendel, S., Matsuyama, S., and Reed, J. C. (1998) *Biochemistry* **37,** 6410–6418
38. Matsuyama, S., Schendel, S., Xie, Z., and Reed, J. (1998) *J. Biol. Chem* **273,** 30995–31001
39. Marzo, I., Brenner, C., Zamzami, N., Jurgensmeier, J. M., Susin, S. A., Vieira, H. L. A., Prevost, M.-C., Xie, Z., Matsuyama, S., Reed, J. C., and Kroemer, G. (1998) *SCIENCE* **281,** 2027–2031
40. Asoh, S., Nishimaki, K., Nanbu-Wakao, R., and Ohta, S. (1998) *J. Biol. Chem.* **273**(18), 11384–11391

41. Lewis, S., Bethell, S., Patel, S., Martinou, J.-C., and Antonsson, B. (1998) *Protein Expr. Purif.* **13,** 120–126
42. Schendel, S., Montal, M., and Reed, J. C. (1998) *Cell Death Differ.* **5**(5)**,** 372–380
43. Muchmore, S. W., Sattler, M., Liang, H., Meadows, R. P., Harlan, J. E., Yoon, H. S., Nettesheim, D., Changs, B. S., Thompson, C. B., Wong, S., Ng, S., and Fesik, S. W. (1996) *NATURE* **381,** 335–341
44. Choe, S., Bennett, M. J., Fujii, G., Curmi, P. M. G., Kantardjieff, K. A., Collier, R. J., and Eisenberg, D. (1992) *NATURE* **357,** 216–221
45. Parker, M. W., Postma, J. P. M., Pattus, F., Tucker, A. D., and Tsernoglou, D. (1992) *J. Mol. Biol.* **224,** 639–657
46. Elkins, P., Bunker, A., Cramer, W. A., and Stauffacher, C. V. (1997) *Structure* **5,** 443–458
47. Weiner, M., Freymann, S., Ghosh, P., and Stround, R. M. (1997) *Nature* **385,** 461–464
48. Minn, A. J., Velez, P., Schendel, S. L., Liang, H., Muchmore, S. W., Fesik, S. W., Fill, M., and Thompson, C. B. (1997) *NATURE* **385,** 353–357
49. Antonsson, B., Conti, F., Ciavatta, A., Montessuit, S., Lewis, S., Martinou, I., Bernasconi, L., Bernard, A., Mermod, J.-J., Mazzei, G., Maundrell, K., Gambale, F., Sadoul, R., and Martinou, J.-C. (1997) *SCIENCE* **277,** 370–372
50. Schlesinger, P., Gross, A., Yin, X.-M., Yamamoto, K., Saito, M., Waksman, G., and Korsmeyer, S. (1997) *Proc. Natl. Acad. Sci. USA* **94,** 11357–11362
51. Montal, M., and Mueller, P. (1972) *Proc. Natl. Acad. Sci. USA* **69**(12)**,** 3561–3566
52. Kagan, B. L., and Sokolov, Y. (1994) *Methods in Enzymology* **235,** 691–705
53. Schendel, S. L., and Cramer, W. A. *Unpublished results*
54. Zakharov, S. D., Heymann, J. B., Zhang, Y.-L., and Cramer, W. A. (1996) *Biophys. J.* **70,** 2774–2783
55. Kawaga, Y., and Racker, E. (1971) *J. Biol. Chem.* **246,** 5477–5487
56. Bartlett, G. R. J. (1959) *J. Biol. Chem.* **234,** 466–472
57. Peterson, A. A., and Cramer, W. A. (1987) *J. Membr. Biol.* **99,** 197–204
58. Cramer, W. A., Heymann, J. B., Schendel, S. L., Deriy, B. N., Cohen, F. S., Elkins, P. A., and Stauffacher, C. V. (1995) *Annu. Rev. Biophys. Biomol. Struct.* **24,** 611–641
59. Sato, T., Hanada, M., Bodrug, S., Irie, S., Iwana, N., Boise, L.H., Thompson, C.B., Golemis, E., Fong, L., Wang, H-G., and Reed, J.C. (1994) *Proc. Natl. Acad. Sci. USA* **91,** 9238–9242
60. Hanada, M., Aimé-Sempé, C., Sato, T., and Reed, J.C. (1995) *J. Biol. Chem.* **270,** 11962–11968
61. Zha, H., Fisk, H., Yaffe, M., Mahajan, N., Herman, B., and Reed, J.C. (1996) *Mol. Cell. Biol.* **16,** 6494–6508
62. Greenhalf, W., Stephan, C., and Chaudhuri, B. (1996) *FEBS Lett.* **380, 169–175**
63. Jurgensmeier, J., Krajewski, S., Armstrong, R., Wilson, G., Oltersdorf, T., Fritz, L., Reed, J., and Ottilie, S. (1997) *Mol.Biol. Cell* **8,** 325–339
64. Ink, B., Zornig, M., Baum, B., Hajibagheri, N., James, C., Chittenden, T., and Evan, G. (1997) *Mol. Biol. Cell* **17**(5), 2468–2474
65. Tao, W., Kurschner, C., and Morgan, J.I. (1997) *J. Biol. Chem.* **272**(24), 15547–15552
66. Kozak, M. (1991) *J. Cell. Biol.* **115,** 887–903
67. Hinnebusch, A. G., and Liebman, S. W. (1991) *in The Molecular and Cellular Biology of the Yeast Saccharomyces: Genome Dynamics, Protein Synthesis, and Energetics (Broach, J. R., Pringle, J. R., and Jones, E. W., Eds.)* **Cold Spring Harbor Laboratory Press, Cold Spring Harbor, NY,** 627–735
68. Mumberg, D., Muller, R., and Funk, M. (1994) *Nucl. Acids. Res.* **22,** 5767–5768
69. Mumberg, D., Muller, R., and Funk, M. (1995) *Gene* **156,** 119-122
70. Lew, D. J., Dulic, V., and Reed, S. I. (1991) *CELL* **66**(6)**,** 1197–1206
71. Toh-e, A., Ueda, Y., Kakimoto, S.-I., and Oshima, Y. (1973) *J. Bacteriol.* **113,** 727–738
72. Sherman, F. (1991) *in Guide to Yeast Genetics and Molecular Biology. (Guthrie, C., and Fink, G. R. Eds.), Academic Press, San Diego.* **194,** 3–21
73. Sherman, F., and Hicks, J. (1991) *in Guide to Yeast Genetics and Molecular Biology. (Guthrie, C., and Fink, G. R. Eds.), Academic Press, San Diego* **194,** 21–38
74. Ito, H., Fududa, Y., Murata, K., and Kimura, A. (1983) *J. Bacteriol.* **153,** 163–168

Methods for detecting proteolysis during apoptosis in intact cells

APRIL L. BLAJESKI and SCOTT H. KAUFMANN

1. Introduction

Over the past five years it has become clear that protease activation is the single most important event that characterizes the process of apoptosis. In the present chapter, we briefly review the evidence that proteases—particularly caspases—play an important role in apoptosis. We then discuss various methods that have been applied to the study of caspases and their precursors, including immunoblotting, activity assays, and affinity labelling. We also describe various polypeptides whose cleavage can be used as a marker for caspase activation *in situ*. Finally, we outline an approach for determining whether a particular cleavage is mediated by caspases, focusing on the potential pitfalls in relying solely upon *in vitro* cleavage assays or inhibitor studies.

2. Involvement of proteases in apoptosis

As described elsewhere in this volume, apoptosis is a morphologically and biochemically distinct form of cell death that involves disassembly of the cell without rupture and leakage of intracellular contents. Four important lines of evidence support a role for caspases in the apoptotic process. First, these proteases have been identified as the enzymes responsible for the proteolysis of numerous nuclear and cytoplasmic proteins that are cleaved during apoptosis (1). Second, proteolytic cleavage of certain caspase zymogens to yield the active enzymes correlates with appearance of the apoptotic phenotype. Third, treatment with low molecular weight peptide-based caspase inhibitors, as well as overexpression of endogenous caspase inhibitors (e.g. inhibitor of apoptosis proteins), prevents apoptosis in cell-free models, in intact cells, and in whole animals (1, 2). Finally, mice containing disruptions in the genes for caspase-3 (3) or caspase-9 (4) have severe defects in brain development, presumably due to the lack of extensive apoptosis that normally occurs during ontogeny.

Table 1. Selected properties of ICE family proteases

New name	Old name	Molecular weight[a]			Preferred small substrates[b]		Commercial sources of antibodies
		Pre	Lg	Sm			
Caspase 1	ICE	45	24 20	14 10	YEVD/X	WEHD/X	Pharmingen, Upstate Biotechnology, Santa Cruz Biotechnology, Oncogene Research Products
Caspase 2	Ich-I_L, NEDD2	48	32 18	14 12	VDVAD/X	DEHD/X	Pharmingen, Upstate Biotechnology, Santa Cruz Biotechnology, Transduction Laboratories, Oncogene
Caspase 3	CPP32, YAMA, Apopain	32	20 17	12	DMQD/X	DEVD/X	Chemicon, Pharmingen, Upstate Biotechnology, Santa Cruz Biotechnology, Oncogene, Transduction Laboratories
Caspase 4	Tx, Ich-2, ICE_rel_II				LEVD/X	(W/L)EHD/X	Pharmingen, Oncogene
Caspase 5	ICE_rel_III, Ty				Unknown	(W/L)EHD/X	
Caspase 6	Mch2	34	21 18	13 11	VEID/X	VEHD/X	Santa Cruz Biotechnology
Caspase 7	Mch3, CMH-1, ICE-LAP3	34	20	12	DEVD/X	DEVD/X	Transduction Laboratories, Santa Cruz Biotechnology
Caspase 8	Mch5, FLICE, MACH	53 55	43 18	12 11	IETD/X	LETD/X	Pharmingen, Santa Cruz Biotechnology
Caspase 9	ICE-LAP6, Mch6	50	37	12	Unknown	LEHD/X	Pharmingen, Santa Cruz Biotechnology
Caspase 10	Mch4, FLICE-2	55	43 17	12	IEAD/X	Unknown	Santa Cruz Biotechnology
Caspase 13	ERICE				Unknown	Unknown	

[a]Pre, Lg and Sm denote the MW of the respective caspase precursor, large subunit and small subunit. The appearance of multiple entries indicates partially processed and fully processed large and small subunits that result from sequential cleavage at the C-terminal end of the large subunit, followed by removal of the linker peptide from the small subunit and the prodomain from the large subunit. A blank in this column indicates that the MW of the processed forms has not been reported.

[b]The left column indicates the preferred substrate specificity reported by Talanian et al. (32), whereas the right column indicates that reported by Thornberry et al. (12).

Although the preceding observations clearly implicate caspases in apoptotic biochemical events, it is important to realize that other proteases can also play a role. Granzyme B, a serine protease present in granules of cytotoxic lymphocytes, can proteolytically activate certain caspase precursors and also directly cleave certain caspase substrates *in situ* and *in vitro* (5), as discussed in Chapter 5. The serine protease A24 and the cysteine protease cathepsin B appear to be activated downstream of caspases in certain model systems (1), raising the possibility that these proteases might be responsible for some of the protein cleavages that occur during apoptosis. Calpains have also been implicated in the apoptotic process (1, 6), as have cathepsin D and the proteosome (2). With the exception of actin, which is thought to be cleaved by calpains in some cell types (7), the substrates of these various proteases during apoptosis remain to be identified. Although the present chapter focuses on caspases, the approach outlined below should be applicable to the study of other proteases and their potential substrates.

3. Caspase nomenclature and classification

The observation that *ced-3*, a gene that is essential for apoptosis in the nematode *Caenorhabditis elegans*, encodes a polypeptide with extensive homology to the cysteine protease interleukin-1β-converting enzyme (ICE) (8) prompted intensive investigation into the potential role of this family of proteases in various models of apoptosis. To date, 13 mammalian members of this family have been described, 11 in humans (1). A list of the known human caspases, along with their old names, molecular weights, and sequence preferences, is found in *Table 1*.

All of these enzymes share several properties. First, they are cysteine-dependent proteases that cleave polypeptides on the carboxyl side of aspartate residues. This activity led to the current designation of these enzymes as 'caspases' (9). Second, each of these enzymes is an $\alpha_2\beta_2$ tetramer consisting of two 17–20 kDa large subunits and two 10–12 kDa small subunits. Finally, each of the caspases is synthesized as a zymogen that encodes a prodomain, a large subunit, and a small subunit. Activation of these zymogens (*Figure 1*) involves proteolytic cleavage between the large and small subunits, followed by removal of the prodomain (1). Caspases themselves and the serine protease granzyme B are currently the only enzymes known to be capableof catalysing this activation process under physiological conditions (10–12).

Based on either sequence homology or function (1), these enzymes can be further divided into at least two broad categories, 'initiator' caspases and 'effector' caspases. Initiator caspases catalyse the cleavage of other caspases, thereby initiating a caspase cascade (1, 13). The initiator caspases typically contain long prodomains that interact with other intracellular components. For example, the prodomains of caspases-8 and -10 contain structural motifs

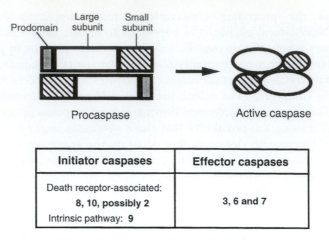

Initiator caspases	Effector caspases
Death receptor-associated: 8, 10, possibly 2 Intrinsic pathway: 9	3, 6 and 7

Figure 1. Schematic representation of caspase activation. As indicated in the text, caspases are synthesized as zymogens that contain three distinct domains. Activation involves cleavage between the large and small subunits followed by removal of the prodomain (1). Several of the caspases that have long prodomains serve as initiator caspases at the apex of protease cascades initiated by death receptors or endogenous signals (1). In contrast, the caspases with short prodomains appear to be responsible for cleavage of most of the substrates undergoing proteolysis during apoptosis.

known as death effector domains (DEDs) that can interact with the homologous domains on adaptor proteins such as FADD/MORT1 (1, 13). Likewise, a caspase recruitment domain (CARD) found in the prodomain of caspase-9 appears to promote interactions with other cytoplasmic components (e.g. the Apaf 1/cytochrome *c* complex) (14, 15). According to one current model, the protein–protein interactions between prodomains and their interacting partners result in the juxtaposition of multiple procaspase molecules. Because these zymogens have low levels of intrinsic catalytic activity, this induced proximity allows one procaspase molecule to activate a neighbouring molecule in an 'autocatalytic' fashion (1).

Once activated, the initiator caspases are then able to cleave, and thereby activate, caspases that contain short prodomains, i.e. procaspases-3, -6, and -7 (1, 13). These so-called effector caspases cleave a variety of substrates throughout the cell (*Table 2*), thereby setting into motion the biochemical and morphological changes that constitute the apoptotic process (1).

4. Detection of procaspases and their cleavage products by immunoblotting

4.1 Theoretical considerations

Because of the intimate involvement of caspases in apoptosis, there has been considerable interest in studying their activation and activity under a variety

Table 2. Polypeptides cleaved by caspases in apoptotic cells[a]

Polypeptides	Cleavage site	Responsible caspase	Sizes reported (kDa)	
			Intact	Fragment(s)[b]
Abundant cytoplasmic proteins				
Gelsolin	DQTD/G	3	83	41, 39
Gas-2	SRVD/G	?	35	31
Fodrin	DETD/S	3	240	150
β-Catenin	?	3	92	65
Cytokeratin 18	VEVD/A	3, 6, 7	45	29, 23
Abundant nuclear proteins				
Lamin A	VEID/N	6	69	47
Lamin B$_1$	VEVD/S	6,?3	67	45
NuMA	?	3,6	230	160, 180
HnRNP proteins C1 and C2	?	3,7	40	~35
70 kDa protein of U1 snRNP	DGPD/G	3	70	40
mdm2	DVPD/C	3, 6,7	100	55
Proteins involved in DNA metabolism and repair[c]				
Poly(ADP-ribose) polymerase	DEVD/G	3, 7, 9	113	89, 24
DNA-PK$_{cs}$	DEVD/N	3	460	150
Replication factor C large subunit	DEVD/G	3	140	87, 53
Topoisomerase I	DDVD/Y	3	100	70
Protein kinases				
Protein kinase Cδ	DMQD/M	3	78	40
Protein kinase Cθ	DEVD/K	3	78	40
Protein kinase C-related kinase 2	DITD/C	3	130	110, 100
Calcium/calmodulin-dependent protein kinase IV	PAPD/A	3	55	38
p21-activated kinase 2	SHVD/G	3, 8	65	36
PITSLRE kinase α2-1	YVPD/S	3	110	60,43
Mst1 kinase	DEMD/S	?	63	34–36
Mst2 kinase	DELD/S	?	63	34
Focal adhesion kinase	DQTD/S	3,7	125	85, 77
	VSWD/S	6		
MEKK-1	DTVD/G	3	200	~100, 120
Wee1 kinase	?	3, 7, 8	~80	~60
Other proteins involved in signal transduction and gene expression				
Pro-interleukin-1β	FEAD/G	1	31	28
	YVHD/A			17.5
Pro-interleukin-16	SSTD/S	3	50	20
Pro-interleukin-18	LESD/Y	1,3	24	18, 16, 15
Ras GTPase activating protein	DTVD/G	3	120	~80, ~65
D4-GDP dissociation inhibitor	DELD/S	3	28	23
Protein phosphatase 2A subunit Aα	DEQD/S	3	65	42
Cytosolic phospholipase A$_2$	DELD/T	3	100	70

219

Table 2. *Continued*

Polypeptides	Cleavage site	Responsible caspase	Sizes reported (kDa) Intact	Fragment(s)[b]
Stat1	MELD/G	3	91, 84	81
NF-κB p65	?	3	65	55, 10
NF-κB p50	?	3	50	35, 15
IκB	DRHD/S	3	40	36
Sterol response element binding protein-1	SEPD/S	3,7	~140	60–70
Sterol response element binding protein-2	DEPD/S	3,7	~140	60–70
Proteins involved in regulation of cell cycle and proliferation[d]				
p21[waf1/cip1]	DHVD/L	3, 7	21	1416
p27[kip1]	DPSD/S	3, 7	27	22
Rb retinoblastoma protein	DEAD/G	3	105	100
CDC 27	?	3	97	~60, ~40
Proteins involved in human genetic diseases				
Dentatorubral pallidalysian atrophy protein	DSLD/G	3	160	145
Presenilin-1, C-terminal fragment	ARQD/S	?	23	10–14
Presenilin-2, C-terminal fragment	DSYD/S	3	25	20
Apoptotic regulatory proteins[e]				
Bcl-2	DAGD/V	?	26	23
Bcl-X[L]	HLAD/S	Not 3	28	16
FLIP[L]	LEVD/G	3, 8, 10	55	43
BID	LQTD/G	8	25	16
BAX	FIQD/R	?	21	18
ICAD/DFF45	DEPD/S	3	45	30,11

[a]For a list of proposed biological functions of cleavage, as well as a comprehensive list of references, see ref. 1.
[b]Estimated masses of caspase-generated fragments are based on mobility in SDS–polyacrylamide gels as reported by authors of the original reports describing cleavage in intact cells. ~ sign indicates our estimate based on published gels in cases where fragment sizes were not reported. Fragments generated by other proteases (e.g. calpains) are not listed.
[c]See also the HnRNP particle proteins C1 and C2 as well as the U1 snRNP particle 70 kDa polypeptide in other parts of this table.
[d]See also Wee1 kinase, mdm2 protein, and replication factor C in other parts of this table.
[e]In addition, all of the procaspases are caspase substrates.

of conditions. Immunoblotting provides one potential method for detecting active caspases within cells. In principle, antibodies generated against neo-epitopes (i.e. epitopes that are generated during the proteolytic activation of caspases) could be utilized to selectively detect active caspases on immunoblots. Such antibodies are not generally available. Instead, most commercially available anti-caspase antibodies (*Table 1*) have been raised against epitopes present in either the large or small subunits of the active enzymes. These antibodies should theoretically recognize the corresponding zymogen and active caspase with equal affinities.

In applying these antibodies to the study of cells undergoing apoptosis, one might expect that decreases in zymogen levels would be accompanied by stoichiometric increases in active caspase species. This is not usually observed. Although the signals for certain procaspases decrease or disappear during apoptosis, suggesting that these zymogens have been activated, a corresponding increase in the large or small subunit is not always detected by immunoblotting. One possible explanation for this phenomenon is a rapid degradation of caspase species once they are activated. Although this explanation remains to be rigorously tested, the detection of active caspases by immunoblotting remains difficult in many model systems. Accordingly, the disappearance of the procaspase signal is commonly accepted as evidence of caspase activation.

Despite the widespread use of anti-caspase antibodies to study caspase activation, published data do not indicate whether the caspase zymogens are abundant or rare polypeptide species. Data in *Figure 2* indicate that 3×10^5 K562 human leukaemia cells contain ~ 10 ng of procaspase-3 and procaspase-8, with 10-fold lower levels of procaspase-6. This corresponds to $\sim 6 \times 10^5$, 4×10^5, and 6×10^4 molecules of procaspase-3, -8, and -6, respectively, per cell. Based on a cell diameter of 15 μm, this translates into a concentration of

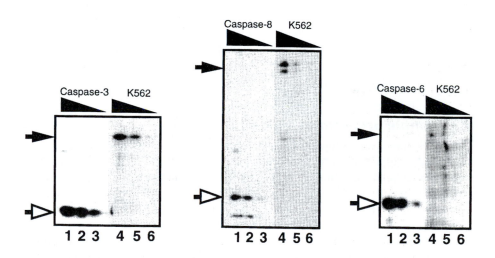

Figure 2. Estimation of procaspase levels in K562 human leukaemia cells. Samples containing 20, 10, or 5 ng of the indicated purified recombinant caspase (lanes 1–3, respectively) or 3×10^5, 1.5×10^5, and 0.75×10^5 K562 cells (lanes 4–6, respectively) were subjected to SDS–PAGE, followed by immunoblotting with antibodies raised against the indicated caspase. Closed arrowhead, caspase zymogen. Two alternatively spliced variants of caspase-8 are present in K562 cells. Open arrowhead, large subunit of active caspase.

~3 μg/ml (100 nM) for procaspase-3 if the zymogen is uniformly distributed throughout the cell. Additional experiments in our laboratory have revealed that procaspase-3 levels in a panel of 60 different human cancer cell lines range from undetectable (in a cell line with a caspase-3 gene rearrangement) to four times as high as K562 cells. Procaspase-8 levels in the same cell lines range from 0.5 to six times as high as K562 cells.

4.2 Practical considerations in immunoblotting for caspases

In designing immunoblotting experiments, it is important to remember that caspases are themselves protease substrates. Zapata *et al.* (16) have reported that lysis of lymphoid cells in neutral detergents such as Triton X-100 can lead to artefactual caspase activation by granzyme B present in cytotoxic granules. Other serine proteases can also activate certain caspases under cell-free conditions (17). It is, therefore, important that cells be solubilized under conditions that prevent procaspase cleavage and activation after cell lysis. We favour the lysis of cells under strongly denaturing conditions [e.g., in 2% sodium dodecyl sulfate (SDS) containing 4 M urea and reducing agent (18) or in 6 M guanidine hydrochloride containing reducing agent (19)]. If the latter buffer is used, sample preparation involves more steps (dialysis to remove the positively charged guanidium ions and replace them with dodecyl sulfate as described in ref. 19), but the chances of artefactual proteolysis are diminished by the rapidity with which guanidine denatures polypeptides. Protein levels are then quantified by one of several methods, including the Bradford assay (20) or the bicinchoninic acid (BCA) assay (21).

4.3 Sample preparation and immunoblotting

Once cells have been subjected to a pro-apoptotic stimulus, protein should be harvested from a minimum of 10^6 cells per data point. After the protein in the samples has been quantitated, equal amounts should be loaded into wells of an SDS–polyacrylamide gel containing a 5–15% (w/v) acrylamide gradient. This will allow detection of the caspases, which are low molecular weight polypeptides, as well as their putative substrates on the same membrane. After separation, the polypeptides are electrophoretically transferred to either nitrocellulose or polyvinylidene fluoride (PVDF) membranes and probed as described in *Protocol 1*. *Figure 2* contains results obtained using this protocol.

Protocol 1. Detection of caspases by immunoblotting

Reagents

- chemiluminescent reagents suitable for use with peroxidase-coupled secondary antibodies include ECL from Amersham or SuperSignal ULTRA® from Pierce
- sources of anti-caspase antibodies are listed in *Table 1*

Method

1. After transfer, block all of the non-specific binding sites on the membrane with TSM* [10% dry milk, 10 mM Tris–HCl (pH 7.4 at room temp), 150 mM NaCl, 100 U/ml penicillin G, 100 μg/ml streptomycin, 1 mM sodium azide] for at least 1 h. This blocking step should be performed immediately after the transfer if PVDF membranes are used. Nitrocellulose membranes can be dried before blocking.

2. Dilute the primary antibody in TSM* according to the manufacturer's instructions and incubate with the membrane overnight at room temperature. Although it is possible to incubate the membrane with primary antibody for shorter periods of time (e.g. 1 h), this often requires a higher concentration of antibody without any improvement in the signal-to-noise ratio.

3. Remove the primary antibody and store it at 4°C for subsequent use. Wash the membrane in PBS (137 mM NaCl, 2.7 mM KCl, 1.5 mM KH_2PO_4, 8 mM Na_2HPO_4, pH 7.4) containing 0.05% (w/v) Tween 20 for 3×15 min. If there is extensive background on the film after an initial immunoblotting experiment, it may be helpful to use a wash buffer of increased stringency (e.g. PBS–0.05% Tween 20 supplemented with 2 M urea or additional NaCl) in subsequent blotting experiments.

4. Wash the membrane with PBS for 2×5 min.

5. Dilute the secondary antibody in PBS containing 3% (w/v) dry milk according to the supplier's instructions. Add antibody to the membrane and incubate for 1 h at room temperature.

6. Remove the secondary antibody (discard) and wash the membrane with PBS–0.05% Tween 20 for 2×5 min, 2×15 min, and 2×5 min.

7. Add detection reagent. If peroxidase-coupled secondary antibodies are used, prepare luminescent reagents and add to the membrane according to the manufacturer's instructions. This step will be omitted if radioactive secondary antibodies are used.

8. Expose the membrane to X-ray film for an appropriate length of time.

5. Cleavage of caspase targets

Since the initial description of selective protein degradation during apoptosis (22), cleavage of a number of cellular polypeptides has been demonstrated in one or more model systems (for recent reviews, see refs 1 and 23). It is important to stress at the outset, however, that the vast majority of species detected by one- or two-dimensional gel electrophoresis do not change during the course of apoptosis (1, 22), indicating that the proteolytic cleavages are selective.

There are three major goals in studying the cleavage of protease targets during apoptosis: (a) to establish that proteases are activated *in situ*, (b) to study the fate of a particular target during apoptosis, and (c) to determine the biological effects of protease activation. Each of these goals requires a somewhat different approach.

5.1 Establishing that proteases are activated *in situ*

As indicated above, blotting with anti-caspase antibodies can be utilized to establish that caspase zymogens are cleaved during apoptosis. Unfortunately, this approach does not establish whether the resulting caspase fragments, if detectable, have enzymatic activity or not. One way to address this question is to determine whether caspase substrates have been cleaved *in situ*.

Table 2 lists polypeptides that are known to be cleaved by caspases during apoptosis. The criteria used to establish that each of these polypeptides is cleaved by caspases include the following: (a) caspases cleave these poly-peptides *in vitro*; (b) the fragments generated during apoptosis correspond to the fragments generated by caspases *in vitro*; and (c) replacement of the P_1 aspartate residue results in polypeptide species that cannot be cleaved *in situ* during apoptosis. Each of the polypeptides in *Table 2* meets the first two criteria and many meet the third as well. Antibodies to several of these polypeptides, including poly(ADP-ribose) polymerase, lamin A, lamin B_1, p21[waf1/cip1], focal adhesion kinase (FAK), and ICAD/DFF45, are commercially available. An immunoblot demonstrating cleavage of one or more of these substrates to their signature fragments (*Table 2*) should be sufficient to establish that caspases have been activated *in situ*. Because caspase targets, like the caspases themselves, can be cleaved artefactually during sample preparation, it is important to solubilize cells under appropriate conditions for this analysis (see Section 4.2).

5.2 Establishing that a polypeptide is cleaved by caspases

Table 3 contains a list of additional polypeptides that are reportedly cleaved as cells undergo apoptosis. These polypeptides do not currently meet the criteria listed in the preceding section. To establish that one of these poly-peptides (or any other polypeptide of interest) is cleaved by caspases during

Table 3. Other polypeptides cleaved during apoptosis[a]

Abundant cytoplasmic polypeptides
Plakoglobin
Adenomatous polyposis coli gene product
Endosome fusion protein rabaptin-5
Plasminogen activator inhibitor-2
Vimentin

Abundant nuclear polypeptides
Lamin B receptor
Chromosome scaffold-binding protein SAF-A

Polypeptides involved in DNA metabolism and repair
Human RAD51
MCM3
DNA topoisomerase II
RNA polymerase I upstream binding factor UBF

Protein kinases
Raf1
Akt1
c-src

Other polypeptides involved in signal transduction pathways
Adapters Cb1 and Cb1-b
Phospholipase C-γ-1
14-3-3β and 14-3-3ϵ

Polypeptides involved in regulation of cell cycle and proliferation
CDC27

Polypeptides involved in human genetic diseases[b]
Huntingtin
The dentatorubropallidoluysian atrophy protein atrophin-1
The spinocerebellar atrophy type 3 protein ataxin-3
The androgen receptor (altered in spinal bulbar muscular atrophy)
Amyloid precursor protein (APP)

[a]In addition to the polypeptides listed in *Table 2*, these polypeptides are reportedly cleaved during apoptosis. It is important to realize that the involvement of caspases in these cleavages remains to be established. Moreover, cleavages of some of these polypeptides occur late in apoptosis and are incomplete. For comprehensive list of references, see ref. 1.
[b]Cleavages of these polypeptides by purified caspases have been observed *in vitro*. Current evidence suggests that the cleavage of huntingtin occurs at a different site in intact cells (reviewed in ref. 1). Cleavages of the other polypeptides have not been demonstrated in intact cells.

apoptosis requires more than the simple demonstration that cleavage occurs in apoptotic cells, for any one of the many proteases that are activated during apoptosis (see Section 2) could, in principle, be responsible for a particular cleavage. Moreover, because some of these proteases are clearly activated downstream of caspases (24, 25), the demonstration that a 'caspase inhibitor' inhibits cleavage also does not establish that a particular polypeptide is a caspase substrate.

The first step in establishing that a particular polypeptide is a caspase substrate *in situ* is to show that caspases can cleave the polypeptide *in vitro*. The availability of purified recombinant caspases-3, -6, -7, and -8 (e.g. Pharmingen, San Diego, CA) has simplified this process. The polypeptide of interest (or a cell fraction containing the polypeptide) should be incubated with the active caspase under cell-free conditions that are compatible with caspase action (e.g. in the presence of 2–10 mM dithiothreitol at neutral pH). Agents that might inhibit caspases, including sulfhydryl alkylating agents (e.g. iodoacetamide) and peptidomimetic chloromethylketones such as TPCK and TLCK (26), should be avoided. If possible, controls should be included to demonstrate that the caspase is active under the conditions used to cleave the potential substrate. After incubation, the polypeptide of interest should be assayed for biological activity (e.g. enzyme activity) and subjected to SDS–PAGE followed by analysis (e.g. staining of the gel or immunoblotting) to determine whether cleavage has occurred. The actual cleavage site(s) can be determined by subjecting the carboxyl terminal fragment(s) to automated Edman degradation (27).

Detection of cleavage by caspases *in vitro* does not establish that the same cleavage occurs *in situ*. At the very least, it is important to demonstrate that fragments generated in intact cells co-migrate with fragments generated by caspases *in vitro*. To take a notable example, several groups reported that actin could be cleaved by apoptotic cell lysates or purified caspases *in vitro* (28, 29). Subsequent analysis, however, suggested that actin remains intact in many cell systems during apoptosis (30). In other cell systems, it appears that calpains, not caspases, mediate the apoptotic cleavage of actin (7).

Although the co-migration of cleavage products generated *in vitro* and *in situ* provides strong evidence that a caspase(s) mediates the cleavage of a particular polypeptide during apoptosis, mutagenesis studies can strengthen this point even further. When lamins mutated at putative P1 aspartates were introduced into tissue culture cells, the mutated polypeptides resisted cleavage during subsequent apoptosis (31), providing convincing support for the view that the cleavage site had been successfully determined. In addition, cells containing the non-cleavable lamins exhibited a delay in apoptosis, raising the possibility that lamin cleavage facilitates, but is not absolutely required for, subsequent apoptotic changes in nuclear morphology.

6. Assays of caspase activity

The demonstration that certain polypeptides (including procaspases) have been cleaved to their signature fragments (*Tables 1* and *2*) provides presumptive evidence that caspases might have been activated. Nonetheless, the possibility that other proteases could catalyse cleavage of a particular polypeptide to fragments of similar size must always be kept in mind. Accordingly, investigators commonly bolster the argument that caspases have been activated by assaying directly for caspase activity.

6.1 Theoretical considerations of caspase activity assays

Analysis of sites that are cleaved in various substrate polypeptides (*Table 2*) facilitated the development of rapid fluorogenic and chromogenic assays for caspase activity. Because the caspases have overlapping cleavage site specificities (*Table 1*), these assays are not specific for individual caspases. This potential problem is best illustrated by the data of Talanian *et al.* (32) who analysed the K_m and k_{cat} of purified caspases acting on a variety of low molecular weight substrates. This analysis revealed that caspases-3 and -7 preferred the *p*-nitroaniline (pNA)-coupled peptide Ac-DEVD-pNA (K_m = 11 or 12 µM, respectively); caspase-1 preferred Ac-YEVD-pNA (K_m = 7.3 µM); and caspase-6 preferred Ac-VEID-pNA (K_m = 30 µM). These preferences were not, however, absolute. Caspase-3, for example, displayed a k_{cat}/K_m (a measure of catalytic efficiency) of 21.8×10^4 for Ac-DEVD-pNA, 6.1×10^4 for Ac-VEID-pNA, and 3.9×10^4 for Ac-YEVD-pNA, indicating that efficiencies at cleaving the various substrates differed by only a factor of 5. Aside from the inefficient cleavage of Ac-DEVD-pNA by caspase-6, the other caspases showed similar promiscuity. Accordingly, the cleavage of small peptide-derived substrates in cell extracts is difficult to attribute to individual caspases.

In addition to tetrapeptide or pentapeptide substrates, *in vitro* translated proteins have also been used to assay caspase activity *in vitro* (33–35). The work of Andrade *et al.* (5) has demonstrated the feasibility of measuring the K_m and k_{cat} using these more relevant macromolecular substrates. Recent data suggest that caspase cleavage preferences derived solely from the study of tetrapeptide-based substrates might not accurately predict cleavage sites in substrate polypeptides (36), raising the possibility that macromolecular substrates might still have an important role in the study of caspases. On the other hand, the disadvantages of radiolabelled macromolecular substrates include the need to perform both *in vitro* transcription/translation reactions to generate substrates and SDS–PAGE followed by fluorography to analyse results. In short, this labour-intensive approach is not suitable for high throughput screening. Accordingly, we provide a protocol for a tetrapeptide-based assay.

6.2 Measuring caspase activity in cell lysates

The protocol for measuring caspase activity in cell lysates can be divided into two sections, the first involving the preparation of subcellular fractions and the second measuring the cleavage of various substrates. It is critical that the cell lysis procedure not allow inadvertent caspase activation. To avoid release of lysosomal and other granule proteases, which could conceivably activate caspases after cell lysis (16, 17), we prefer a lysis procedure that avoids detergents and thereby allows removal of sequestered proteases by differential centrifugation.

Protocol 2. Measurement of caspase activity using fluorogenic substrates

Reagents

- procedures for the synthesis of potential substrates are described in detail in ref. 18
- free pNA, 7-amino-4-trifluoromethylcouarin (AFC), or 7-amino-4-methylcoumarin (AMC) to construct standard curves are available from Sigma
- chromogenic and fluorogenic caspase substrates are available from a number of suppliers, including Bachem Bioscience, Biomol, Calbiochem, and Enzyme Systems Products

Method

Part A. Preparation of cytosol or other subcellular fractions (18)

1. Harvest adherent cells by trypsinization followed by sedimentation at 200 g for 10 min. Wash the sample twice in ice-cold PBS. All further steps are performed at 4°C unless otherwise indicated.

2. Resuspend the pellet in a small volume of buffer A [25 mM Hepes (pH 7.5 at 4°C), 5 mM MgCl$_2$, 1 mM EGTA supplemented immediately before use with 1 mM α-phenylmethylsulfonyl fluoride (PMSF), 1 mM dithiothreitol (DTT), 10 μg/ml pepstatin A, and 10 μg/ml leupeptin]. Typically, 1 ml of buffer A is added to 1–3 × 10^8 cells. This yields a cytosolic extract with a protein concentration ranging from 4 to 8 mg/ml.

3. Incubate the sample for 20 minutes, then lyse the cells with 20–30 strokes in a tight-fitting Dounce homogenizer. Mix 3–5 μl of homogenate with an equal volume of 0.4% (w/v) trypan blue and examine by light microscopy. Continue homogenization until >95% of the cells stain blue.

4. Remove the nuclei by sedimentation (800 g for 10 min or 16 000 g for 3 min).

5. Supplement the supernatant with EDTA to a final concentration of 5 mM. Prepare cytosol from the post-nuclear supernatant by sedimentation at 280 000 g_{max} for 60 min in a Beckman TL-100 ultracentrifuge. Other fractions can be likewise be purified by differential sedimentation (37).

6. Freeze the cytosol (280 000 g supernatant) in 50 μl aliquots at –70°C. Control experiments from our laboratory have indicated that activity capable of cleaving Ac-DEVD-AFC is stable for at least 3 months at –70°C.

Part B. Fluorogenic or chromogenic substrate cleavage assay

1. Determine the protein concentration of the cytosol or other subcellular fraction using the method of Bradford (20) or Smith *et al*. (21).

2. Thaw aliquots of cytosol or subcellular fractions containing 50 µg of cytosolic protein at 4°C. Dilute to 50 µl with ice-cold buffer A containing 5 mM EDTA.

3. Dilute sample with 225 µl of freshly prepared buffer B [25 mM Hepes (pH 7.5), 0.1% (w/v) 3-[(3-cholamidopropyl)dimethylammonio]-1-propanesulfonate, 10 mM DTT, 100 U/ml aprotinin, 1 mM PMSF] containing 100 µM substrate.

4. Incubate the reaction at 37°C. Continuous monitoring of fluorochrome release can be utilized to examine the kinetics of product release and/or the kinetics of enzyme inhibition if suitable equipment is available (38). Alternatively, the reaction can be run as an end-point assay (18). In that case, add 1.225 ml ice-cold buffer B at a fixed time point to stop the reaction.

5. Incubate negative control reactions, containing 50 µl of buffer A and 225 µl of buffer B at 37°C and then dilute them with 1.225 ml ice-cold buffer B in parallel with experimental samples.

6. Measure the fluorescence in a fluorometer using an excitation wavelength of 360 nm and an emission wavelength of 475 nm. The absolute amount of fluorochrome released is determined by measuring the fluorescence of a panel of standards containing various amounts of the liberated fluorophore [e.g. 0–1500 pmoles AFC].

When setting up this assay, it is important to use an appropriate substrate concentration. If the substrate concentration is below the K_m for the enzyme, changes in the amount of product released might be due to either changes in the number of active enzyme molecules (altered v_{max}) or the affinity for the substrate (altered K_m). The common interpretation that changes in the amount of product released correspond to changes in the number of active enzyme molecules is only valid when the reaction is run under conditions where the substrate is saturating.

When the assay is run as an end-point assay, it is important to confirm that product release is a linear function of the length of incubation. Important causes of non-linearity include substrate exhaustion and enzyme instability. If extracts are prepared as described above, the assay is linear for up to 4 h under the specified conditions.

Although numerous substrates can be utilized for this assay, peptide derivatives of AMC and AFC are most commonly employed because of the high quantum yield of the liberated fluorophore. Colorimetric assays utilizing chromogenic substrates (e.g. a peptide linked to pNA) have also been developed. These colorimetric assays can easily be adapted to microtitre plate readers, permitting their use in high throughput screening.

7. Detection of active caspases by affinity labelling

7.1 Theoretical considerations in affinity labelling

Affinity labelling provides an alternative approach for demonstrating that caspases have been activated. This approach is based on the observation that the catalytic cysteines in active caspases can be covalently modified by reaction with a variety of agents (reviewed in ref. 39). Reactions with aldehydes and nitriles are reversible, whereas reactions with chloromethyl, fluoromethyl, or acyloxymethyl ketones are irreversible. Coupling of these reactive groups to suitable peptides provides potential enzyme inhibitors as well as affinity labelling reagents.

Two reagents currently in widespread use, N-(acetyltyrosinylvalinyl-N^{ε}-biotinyllysyl) aspartic acid [(2,6-dimethylbenzoyl)oxy]methyl ketone [*Figure 3*; abbreviated to YVK(bio)D-aomk] and N-(N^{α}-benzyloxycarbonylglutamyl-N^{ε}-biotinyllysyl)aspartic acid [(2,6-dimethylbenzoyl)oxy]methyl ketone [abbreviated to Z-EK(bio)D-aomk], will be discussed to illustrate essential features of the affinity labelling reagents. These molecules contain peptide sequences that direct them to caspases, acyloxymethyl ketone moieties to permit covalent coupling to the caspase active site, and biotin groups to allow detection of polypeptides that have been derivatized. Each of these features plays an important role in determining the properties of these reagents.

The presence of the polypeptide moiety enhances the affinity of each reagent for certain proteases. YVK(bio)D-aomk contains a YVKD polypeptide that preferentially binds to caspase-1. Z-EK(bio)D-aomk contains the E-X-D motif found in many polypeptides at sites cleaved by caspases-3, -6, and -7. At least four amino acids (or amino acid-like moieties) to the amino terminal side of the reactive group are required to achieve relatively high affinity to caspases (34, 38). In both YVK(bio)D-aomk and Z-EK(bio)D-aomk, the P_1 amino acid is an aspartate, a group that is absolutely required for binding to caspases (1). Because the P_2 and P_3 side chains point away from the active site of the enzyme, the bulky biotin moiety is tolerated in the P_2 position (1, 40). The group in the P_4 position, on the other hand, is tightly bound to the enzyme. The corresponding subsite on caspase-3 is positively charged and prefers acidic amino acids, whereas the subsite on caspase-1 accommodates bulkier groups like tyrosine. As described below, however, this selectivity is relative rather than absolute.

The acyloxymethylketone group is employed for covalent modification of the caspases because of its somewhat lower reactivity (and presumed greater selectivity) than chloromethyl or fluoromethyl ketones (39, 40). When working with purified caspases, where selectivity is less of an issue, it might be

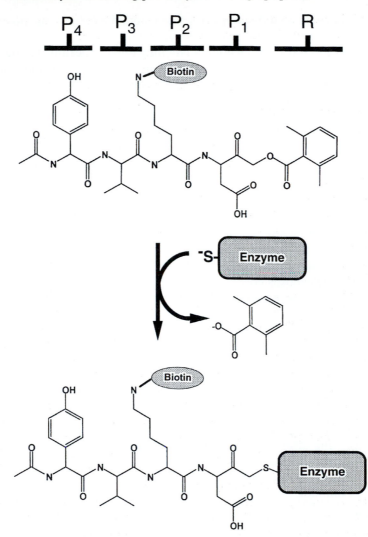

Figure 3. Schematic representation of the chemical reaction involved in affinity labelling of active caspases by YVK(bio)D-aomk (40, 42). Above, the chemical structure for YVK(bio)D-aomk, the locations of the leaving group R, and P1–P4 residues are indicated.

possible to use affinity labelling reagents with the more reactive groups. In whole-cell lysates, however, chloromethyl ketones exhibit relatively promis-cuous reactivity with cysteine proteases (41) in addition to their well-established, but more selective, reactivity with serine proteases. Likewise, fluoromethyl ketones are used to derivatize a variety of sulfhydryl proteases. Because of the widespread reactivity of these types of molecules with polypeptides containing activated cysteine residues, the possibility must be borne in mind that peptide-coupled acyloxymethyl ketones can potentially react with other enzymes, not just caspases.

The presence of biotin in the affinity labelling reagent permits the detection of polypeptides that have been covalently modified (e.g. by blotting with streptavidin). Unfortunately, endogenous biotinylated polypeptides will also be labelled by streptavidin (18), limiting the usefulness of this approach in immunohistochemical studies. Reagents with different detection groups remain to be synthesized.

7.2 Practical considerations in the use of affinity labelling reagents

Despite the original intent to design affinity labelling reagents with some selectivity, YVK(bio)D-aomk and Z-EK(bio)D-aomk lack selectivity for their originally intended caspase targets under the conditions commonly utilized. YVK(bio)D-aomk has been successfully utilized to label the caspases that are present in extracts from apoptotic cells, including the proteases that cleave poly(ADP-ribose) polymerase and the lamins (42); and Z-EK(bio)D-aomk readily reacts with recombinant caspases-1 and -2 as well as caspases-3 and -6 (18). Although these reagents might show some selectivity for various caspases during brief labelling at low ligand concentrations, prolonged incubation with high concentrations of these reagents easily compensates for poor affinity, as demonstrated by Margolin *et al.* (34).

7.3 Labelling and detecting enzymatically active caspases

The overall approach to the use of these reagents involves the preparation of whole-cell lysates or subcellular fractions from control and apoptotic cells, reaction with the affinity label under conditions where caspases are active, separation of polypeptides in one- or two-dimensional polyacrylamide gels, transfer to nitrocellulose, and reaction with peroxidase-coupled streptavidin followed by luminol.

Protocol 3. Detection of active caspases by affinity labelling

Reagents

- affinity labelling reagents are available from a limited number of suppliers, including   Bachem Bioscience, Calbiochem, Enzyme Systems Products, and the Peptide Institute

Method

1. Each sample should contain ~1 × 10^8 cells. At the very least, an experiment should contain cells treated with a pro-apoptotic stimulus and control (non-apoptotic cells).

2. After the induction of apoptosis is complete, wash cells twice in ice-cold serum-free isotonic buffer (e.g. PBS).

---

3. Isolate cytosol or other subcellular fractions using the techniques described in *Protocol 2* and store at –70°C in buffer A supplemented with 5 mM EDTA.

4. Determine protein concentration in an aliquot of the subcellular fraction using an assay that is not affected by reagents in the EDTA-supplemented buffer A [e.g. the bichinconinic acid method (21)].

5. Perform affinity labelling by reacting 30–50 µg of cytosolic protein for 1 h at 20–22°C with 1 µM Z-EK(bio)D-aomk. The Z-EK(bio)D-aomk is conveniently prepared as a 25 mM stock in DMSO and stored in small aliquots at –80°C.

6. To stop the reaction, dilute the sample with 0.5 vol. 3 × concentrated SDS sample buffer [0.15 M Tris–HCl (pH 6.8 at 20°C), 45% (w/v) sucrose, 6 mM EDTA, 9% (w/v) SDS, 0.03% (w/v) bromophenol blue, 10% (v/v) β-mercaptoethanol] and heat to 95–100°C for 3 min.

7. Subject the samples to one-dimensional SDS–PAGE in 16% (w/v) polyacrylamide gels. Alternatively, subject the samples to two-dimensional isoelectric focusing/SDS–PAGE. For the latter procedure, our collaborators have obtained highly reproducible results using pre-cast Immobiline gels (pH 4–7, Pharmacia, Upsalla, Sweden) for the first dimension.

8. After separation, transfer polypeptides to nitrocellulose.

9. Block unoccupied protein-binding sites with PBS containing 0.1% (v/v) Tween-20 (PBS-T) and 5% (w/v) non-fat dry milk.

10. Visualize the labelled polypeptides by reacting with peroxidase-coupled streptavidin in PBS-T for 3–4 h at 20–22°C. Wash blots with PBS-T for 1 × 15 min and 4 × 5 min to remove unbound peroxidase-coupled streptavidin.

11. Detect the bound peroxidase using enhanced chemiluminescent detection.

12. Recombinant caspases expressed in Sf9 cells (18) can be subjected to this analysis in parallel in order to permit identification of the labelled polypeptide species. His_6-tagged caspases (e.g. commercially available from Pharmingen, San Diego, CA) should not be used as standards for two-dimensional analysis because the charged tag will alter their migration.

This assay can detect pg levels of active proteases. Moreover, when utilized in conjunction with known amounts of recombinant caspases, this assay can identify and quantify the individual proteases that are active at particular times during the course of programmed cell death.

Control experiments indicate that cytosol can be successfully labelled after storage at –70°C for at least 3 months. As an alternative to subcellular fractionation, whole-cell lysates prepared by freezing and thawing cells can also be labelled using analogous techniques (43). The caspase labelling pattern in these lysates, however, is more complicated than the pattern observed using cytosol or nuclei, suggesting that specific caspase species might target other subcellular compartments as well.

The recommended 20–22°C temperature for incubation with the affinity label appears to be important. Incubation of cytosol at 37°C results in rapid degradation of active caspase species (P.W. Mesner, Jr, L. Miguel Martins, and S.H. Kaufmann, unpublished observations).

Two types of controls can be performed to distinguish endogenous biotinyl-ated polypeptides from species that become biotinylated as a consequence of reaction with Z-EK(bio)D-aomk. First, as indicated above, control (non-apoptotic) cells can be labelled in parallel with apoptotic cells. Although some non-apoptotic cells contain active caspase-1 (38), there is no evidence for active caspases-3, -6, or -7 in non-apoptotic cells. Secondly, the ability of other caspase inhibitors to prevent labelling can be assessed. For example, pre-incubation of cell lysates with certain peptide chloromethyl or fluoromethyl ketones will completely abolish subsequent labelling of active caspases with Z-EK(bio)D-aomk but will not have any effect on the reaction of streptavidin with endogenous biotinylated polypeptides (44).

It is also possible to combine affinity labelling with subsequent affinity chromatography on avidin–agarose to purify active caspases up to 100 000-fold in a single step (33, 38, 43, 45).

8. Studying the biological effects of caspase cleavages

A recent analysis (1) suggests that caspase-mediated cleavages play three important roles during apoptosis. First, they turn off activities required for cell survival, including extracellular matrix-initiated signalling pathways, DNA repair, and RNA synthesis. Second, they turn on activities that lead to cellular disruption. Among these are gelsolin, which disrupts the actin cytoskeleton; the caspase-activated DNase CAD/CPAN/DFF40, which contributes to internucleosomal DNA degradation; and a variety of protein kinases. Third, they diminish the integrity of certain critical structural components of the cell, including cytokeratin 18, fodrin, the lamins, and the nuclear/mitotic apparatus protein (NuMA). Collectively, these cleavages give rise to the biochemical and morphological changes that comprise the process of apoptosis.

Two major approaches have evolved for studying the effects of caspase activation on this process. Each of these is briefly discussed below.

8.1. A molecular approach for analysis of the effects of substrate cleavage

For polypeptides that are thought to be inactivated by caspase-mediated cleavage, expression of non-cleavable isoforms (e.g. in which the P_1 aspartate has been mutated *in vitro* to an alanine) can provide some evidence for the role of cleavage. This is illustrated by the effect of non-cleavable lamins cited above.

For polypeptides that are thought to be activated by caspase-mediated cleavage (e.g. kinases or other enzymes), forced expression of the putative cleavage product can provide some insight into the biological role of the cleavage. For example, caspase-3 cleaves protein kinase Cδ between its regulatory and catalytic domains to yield a C-terminal fragment with constitutive kinase activity (46). Overexpression of this C-terminal fragment induces apoptotic morphological changes in cells, whereas overexpression of the full-length polypeptide does not (47). By itself, this experiment simply demonstrates that the overexpressed C-terminal fragment is toxic to cells. However, when combined with antisense experiments demonstrating that the downregulation of protein kinase Cδ inhibits apoptosis (47), the net result is strong evidence supporting an active role for the protein kinase Cδ fragment in apoptotic changes. A similar approach has been utilized to determine the biological effects of other caspase-generated enzyme fragments (1).

It should be noted, however, that the preceding experiments demonstrate a role for the caspase-generated enzyme fragment during apoptosis but do not elucidate that role. Although caspase activation has been shown to generate a variety of constitutively active kinase fragments (1), the substrates of these kinases remain to be identified; and the manner in which phosphorylation of these substrates contributes to the apoptotic process remains to be clarified. These are questions that are potentially difficult to address using a molecular approach.

8.2. Use of caspase inhibitors: promises and pitfalls

As an alternative to the molecular genetic approach, it is appealing to attempt to inhibit caspase-mediated cleavages using specific inhibitors. In an ideal world, the role of caspase-3, for example, during apoptosis could be analysed by examining the effect of treating cells with a selective caspase-3 inhibitor. Unfortunately, there are several problems with this approach.

First, as indicated above, multiple proteases appear to be activated downstream of the caspases. Hence, the inhibition of a particular cleavage by a caspase-3 inhibitor might reflect cleavage of that polypeptide by caspase-3 or by a protease activated downstream of caspase-3.

Second, currently available caspase inhibitors do not have sufficient specificity to permit this type of analysis. For example, *N*-(acetylaspartylglutamylvalinyl)-3-amino-3-formyl-propionic acid (abbreviated to Ac-DEVD-CHO), which mimics the cleavage site in poly(ADP-ribose) polymerase (27, 33), is a potent inhibitor of caspase-3 [K_i of 0.2–0.35 nM (5)]. However, it is not selective. Ac-DEVD-CHO also inhibits caspase-1 ($K_i = 17$ nM) at concentrations that overlap those required to inhibit caspase-3 family members [$K_i = 31, 1, 0.8, 60,$ and 12 nM for caspases-6–10, respectively (5)]. As a result, it is quite possible that Ac-DEVD-CHO inhibits a wide range of caspases at the 50–100 µM concentrations somtimes utilized *in vitro* and *in vivo* (48). Moreover, recent experiments indicate that inhibitors such as Ac-DEVD-CHO and Z-VAD-fluoromethyl ketone can also inhibit non-caspase proteases (e.g. cathepsin B) at micromolar concentrations (25). This lack of selectivity is even more problematic with the more reactive chloromethyl ketones.

In summary, the currently available inhibitors are not sufficiently selective to permit the study of a particular caspase. On the contrary, recent evidence demonstrating that these compounds directly inhibit other classes of proteases, coupled with the observation that non-caspase proteases can be activated downstream of caspases, makes it difficult to support the contention that a particular effect is mediated by caspases solely on the basis of inhibition by these compounds. Because of this lack of specificity, investigators have found it necessary to turn to gene disruption experiments (reviewed in ref. 1) to assess the effect of selectively abolishing the activity of individual caspases.

Acknowledgements

We apologize to the numerous authors whose seminal contributions to this field could not be cited because of space limitations. Work in our laboratory is supported by R01 CA69008 and a predoctoral fellowship (to A.L.B.) from the Mayo Foundation. We wish to thank numerous colleagues, including Bill Earnshaw, Miguel Martins, Peter W. Mesner, Jr, Tim Kottke, Phyllis Svingen, Guy Poirier, and Greg Gores, for provocative discussions and helpful comments. Data shown in *Figure 2* were provided by Phyllis Svingen using antibodies kindly provided by Peter W. Mesner, Jr and John Reed. The secretarial assistance of Deb Strauss is gratefully acknowledged.

References

1. Earnshaw, W. C., Martins, L. M., and Kaufmann, S. H. (1999). *Ann. Rev. Biochem.*, **68**, 383.
2. Mesner, P. W. and Kaufmann, S. H. (1998). In: *Apoptosis in Neurobiology: Concepts and Methods* (ed. Hannun, Y. A., and Boustany, R.-M.), p. 73, CRC Press, Boca Ratan, FL.
3. Kuida, K., Zheng, T. S., Na, S., Kuan, C., Yang, D., Karasuyama, H., Rakic, P., and Flavell, R. A. (1996). *Nature*, **384**, 368.

4. Kuida, K., Haydar, T. F., Kuan, C.-Y., Gu, Y., Taya, C., Karasuyama, H., Su, M. S.-S., Rakic, P., and Flavell, R. A. (1998). *Cell*, **94**, 325.
5. Andrade, F., Roy, S., Nicholson, D., Thornberry, N., Rosen, A., and Casciola-Rosen, L. (1998). *Immunity*, **8**, 451.
6. Sarin, A., Clerici, M., Blatt, S. P., Hendrix, C. W., Shearer, G. M., and Henkart, P. A. (1994). *J. Immunol.*, **153**, 862.
7. Villa, P. G., Henzel, W. J., Sensenbrenner, M., Henderson, C. E., and Pettmann, B. (1998). *J. Cell Sci.*, **111**, 713.
8. Yuan, J., Shaham, S., Ledoux, S., Ellis, H. M., and Horvitz, H. R. (1993). *Cell*, **75**, 641.
9. Alnemri, E. S., Livingston, D. J., Nicholson, D. W., Salvesen, G., Thornberry, N. A., Wong, W. W., and Yuan, J. (1996). *Cell*, **87**, 171.
10. Greenberg, A. H. (1996). *Adv. Exp. Med. Biol.*, **406**, 219.
11. Pham, C. T. and Ley, T. J. (1997). *Sem. Immunol.*, **9**, 127.
12. Thornberry, N. A., Rano, T. A., Peterson, E. P., Rasper, D. M., Timkey, T., Garcia-Calvo, M., Houtzager, V. M., Nordstrom, P. A., Roy, S., Vaillancourt, J. P., Chapman, K. T., and Nicholson, D. W. (1997). *J. Biol. Chem.*, **272**, 17907.
13. Fraser, A. and Evan, G. (1996). *Cell*, **85**, 781.
14. Li, P., Nijhawan, D., Budihardjo, I., Srinivasula, S. M., Ahmad, M., Alnemri, E. S., and Wang, X. (1997). *Cell*, **91**, 479.
15. Deveraux, Q. L., Roy, N., Stennicke, H. R., Van Arsdale, T., Zhou, Q., Srinivasula, S. M., Alnemri, E. S., Salvesen, G. S., and Reed, J. C. (1998). *EMBO J.*, **17**, 2215.
16. Zapata, J. M., Takahashi, R., Salvesen, G. S., and Reed, J. C. (1998). *J. Biol. Chem.*, **273**, 6916.
17. Zhou, Q. and Salvesen, G. S. (1997). *Biochem. J.*, **324**, 361.
18. Martins, L. M., Kottke, T. J., Mesner, P. W., Basi, G. S., Sinha, S., Frigon, N., Jr, Tatar, E., Tung, J. S., Bryant, K., Takahashi, A., Svingen, P. A., Madden, B. J., McCormick, D. J., Earnshaw, W. C., and Kaufmann, S. H. (1997). *J. Biol. Chem.*, **272**, 7421.
19. Kaufmann, S. H., Svingen, P. A., Gore, S. D., Armstrong, D. K., Cheng, Y.-C., and Rowinsky, E. K. (1997). *Blood*, **89**, 2098.
20. Bradford, M. M. (1976). *Anal. Biochem.*, **72**, 248.
21. Smith, P. K., Krohn, R. I., Hermanson, G. T., Mallia, A. K., Gartner, F. H., Provenzano, M. D., Fujimoto, E. K., Goeke, N. M., Olson, B. J., and Klenk, D. C. (1985). *Anal. Biochem.*, **150**, 76.
22. Kaufmann, S. H. (1989). *Cancer Res.*, **49**, 5870.
23. Tan, X. and Wang, J. Y. (1998). *Trends Cell Biol.*, **8**, 116.
24. Wright, S. C., Schellenberger, U., Wang, H., Wang, Y., and Kinder, D. H. (1998). *Biochem. Biophys. Res. Commun.*, **245**, 797.
25. Faubion, W. A., Guicciardi, M. E., Miyoshi, H., Bronk, S. F., Roberts, P. J., Svingen, P. A., Kaufmann, S. H., and Gores, G. J. (1999). *J. Clin. Invest.* **103**, 137.
26. Fernandes-Alnemri, T., Takahashi, A., Armstrong, R., Krebs, J., Fritz, L., Tomaselli, K. J., Wang, L., Yu, Z., Croce, C. M., Salveson, G., Earnshaw, W. C., Litwack, G., and Alnemri, E. S. (1995). *Cancer Res.*, **55**, 6045.
27. Lazebnik, Y. A., Kaufmann, S. H., Desnoyers, S., Poirier, G. G., and Earnshaw, W. C. (1994). *Nature*, **371**, 346.
28. Kayalar, C., Ord, T., Testa, M. P., Zhong, L. T., and Bredesen, D. E. (1996). *Proc. Natl. Acad. Sci., USA*, **93**, 2234.

29. Mashima, T., Naito, M., Noguchi, K., Miller, D. K., Nicholson, D. W., and Tsuruo, T. (1997). *Oncogene*, **14**, 1007.
30. Song, Q., Wei, T., Lees-Miller, S., Alnemri, E., Watters, D., and Lavin, M. F. (1997). *Proc. Natl. Acad. Sci. USA*, **94**, 157.
31. Rao, L., Perez, D., and White, E. (1996). *J. Cell Biol.*, **135**, 1441.
32. Talanian, R. V., Quinlan, C., Trautz, S., Hackett, M. C., Mankovich, J. A., Banach, D., Ghayur, T., Brady, K. D., and Wong, W. W. (1997). *J. Biol. Chem.*, **272**, 9677.
33. Nicholson, D. W., Ali, A., Thornberry, N. A., Vaillancourt, J. P., Ding, C. K., Gallant, M., Gareau, Y., Griffin, P. R., Labelle, M., and Lazebnik, Y. A. (1995). *Nature*, **376**, 37.
34. Margolin, N., Raybuck, S. A., Wilson, K. P., Chen, W., Fox, T., Gu, Y., and Livingston, D. J. (1997). *J. Biol. Chem.*, **272**, 7223.
35. Liu, X., Zou, H., Slaughter, C., and Wang, X. (1997). *Cell*, **89**, 175.
36. Samejima, K., Svingen, P. A., Gasi, G. S., Kottke, T., Mesner, P. W., Jr, Stewart, L., Champoux, J., Kaufmann, S. H., and Earnshaw, W. C. (1999). *J. Biol. Chem.*, **274**, 4335.
37. Graham, J. M., and Rickwood, D., eds. (1997). Subcellular Fractionation: A Practical Approach. Oxford University Press, Oxford.
38. Thornberry, N. A., Bull, H. G., Calaycay, J. R., Chapman, K. T., Howard, A. D., Kostura, M. J., Miller, D. K., Molineaux, S. M., Weidner, J. R., and Aunins, J. (1992). *Nature*, **356**, 768.
39. Thornberry, N. A. and Molineaux, S. M. (1995). *Protein Sci.*, **4**, 3.
40. Thornberry, N. A., Peterson, E. P., Zhao, J. J., Howard, A. D., Griffin, P. R., and Chapman, K. T. (1994). *Biochemistry*, **33**, 3934.
41. Shaw, E. (1990). *Adv. Enzymol. Relat. Areas Mol. Biol.*, **63**, 271.
42. Takahashi, A., Alnemri, E. S., Lazebnik, Y. A., Fernandes-Alnemri, T., Litwack, G., Moir, R. D., Goldman, R. D., Poirier, G. G., Kaufmann, S. H., and Earnshaw, W. C. (1996). *Proc. Natl. Acad. Sci. USA*, **93**, 8395.
43. Martins, L. M., Kottke, T. J., Kaufmann, S. H., and Earnshaw, W. C. (1998). *Blood*, **92**, 3042.
44. Martins, L. M., Mesner, P. W., Kottke, T. J., Basi, G. S., Sinha, S., Tung, J. S., Svingen, P. A., Madden, B. J., Takahashi, A., McCormick, D. J., Earnshaw, W. C., and Kaufmann, S. H. (1997). *Blood*, **90**, 4283.
45. Faleiro, L., Kobayashi, R., Fearnhead, H., and Lazebnik, Y. (1997). *EMBO J.*, **16**, 2271.
46. Emoto, Y., Manome, Y., Meinhardt, G., Kisaki, H., Kharbanda, S., Robertson, M., Ghayur, T., Wong, W. W., Kamen, R., Weichselbaum, R., and Kufe, D. (1995). *EMBO J.*, **14**, 6148.
47. Ghayur, T., Hugunin, M., Talanian, R. V., Ratnofsky, S., Quinlan, C., Emoto, Y., Pandey, P., Datta, R., Huang, Y., Kharbanda, S., Allen, H., Kamen, R., Wong, W., and Kufe, D. (1996). *J. Exp. Med.*, **184**, 2399.
48. Dubrez, L., Savoy, I., Hamman, A., and Solary, E. (1996). *EMBO J.*, **15**, 5504.

List of suppliers

Amersham

Amersham Pharmacia Biotech UK Ltd, Amersham Place, Little Chalfont, Buckinghamshire HP7 9NA, UK.

Amersham Corporation, 2636 South Clearbrook Drive, Arlington Heights, IL 60005, USA.

Anderman

Anderman and Co. Ltd., 145 London Road, Kingston-Upon-Thames, Surrey KT17 7NH, UK.

ATCC, 10801 University Boulevard, Manassas, VA 20110–2209, USA.

Bachem AG, Hauptstrasse 144, CH-4416, Bubendorf, Switzerland.

BDH Laboratory Supplies, Poole Dorset, BH15 1TD, UK.

Beckman Instruments

Beckman Instruments UK Ltd., Oakley Court, Kingsmead Business Park, London Road, High Wycombe, Bucks HP11 1J4, UK.

Beckman Instruments Inc., PO Box 3100, 2500 Harbor Boulevard, Fullerton, CA 92634, USA.

Becton Dickinson

Becton Dickinson and Co., Between Towns Road, Cowley, Oxford OX4 3LY, UK.

Becton Dickinson and Co., 2 Bridgewater Lane, Lincoln Park, NJ 07035, USA.

Bio

Bio 101 Inc., c/o Statech Scientific Ltd, 61–63 Dudley Street, Luton, Bedfordshire LU2 0HP, UK.

Bio 101 Inc., PO Box 2284, La Jolla, CA 92038–2284, USA.

Bio-Rad Laboratories

Bio-Rad Laboratories Ltd., Bio-Rad House, Maylands Avenue, Hemel Hempstead HP2 7TD, UK.

Bio-Rad Laboratories, Division Headquarters, 3300 Regatta Boulevard, Richmond, CA 94804, USA.

Boehringer Mannheim

Boehringer Mannheim UK (Diagnostics and Biochemicals) Ltd, Bell Lane, Lewes, East Sussex BN17 1LG, UK.

Boehringer Mannheim Corporation, Biochemical Products, 9115 Hague Road, P.O. Box 504 Indianapolis, IN 46250–0414, USA.

Boehringer Mannheim Biochemica, GmbH, Sandhofer Str. 116, Postfach 310120 D-6800 Ma 31, Germany.

Brand Applications, P.O. Box 864 6200 AW Maastricht, The Netherlands.

British Drug Houses (BDH) Ltd, Poole, Dorset, UK.

Difco Laboratories

Difco Laboratories Ltd., P.O. Box 14B, Central Avenue, West Molesey, Surrey KT8 2SE, UK.

Difco Laboratories, P.O. Box 331058, Detroit, MI 48232–7058, USA.

Du Pont

Dupont (UK) Ltd., Industrial Products Division, Wedgwood Way, Stevenage, Herts, SG1 4Q, UK.

Du Pont Co. (Biotechnology Systems Division), P.O. Box 80024, Wilmington, DE 19880–002, USA.

Enzyme Systems Products, 6497 Sierra Lane, Dublin, CA 94568, USA.

European Collection of Animal Cell Culture, Division of Biologics, PHLS Centre for Applied Microbiology and Research, Porton Down, Salisbury, Wilts SP4 0JG, UK.

Falcon (Falcon is a registered trademark of Becton Dickinson and Co.).

Fisher Scientific Co., 711 Forbest Avenue, Pittsburgh, PA 15219–4785, USA.

Flow Laboratories, Woodcock Hill, Harefield Road, Rickmansworth, Herts. WD3 1PQ, UK.

Fluka

Fluka-Chemie AG, CH-9470, Buchs, Switzerland.

Fluka Chemicals Ltd., The Old Brickyard, New Road, Gillingham, Dorset SP8 4JL, UK.

Gibco BRL

Gibco BRL (Life Technologies Ltd.), Trident House, Renfrew Road, Paisley PA3 4EF, UK.

Gibco BRL (Life Technologies Inc.), 3175 Staler Road, Grand Island, NY 14072–0068, USA.

Arnold R. Horwell, 73 Maygrove Road, West Hampstead, London NW6 2BP, UK.

Hybaid

Hybaid Ltd., 111–113 Waldegrave Road, Teddington, Middlesex TW11 8LL, UK.

Hybaid, National Labnet Corporation, P.O. Box 841, Woodbridge, NJ. 07095, USA.

HyClone Laboratories 1725 South HyClone Road, Logan, UT 84321, USA.

ICN Pharmaceuticals Inc., 330 Hyland Avenue, Costa Mesa, CA 92626, USA.

International Biotechnologies Inc., 25 Science Park, New Haven, Connecticut 06535, USA.

Invitrogen Corporation

Invitrogen Corporation 3985 B Sorrenton Valley Building, San Diego, CA. 92121, USA.

Invitrogen Corporation c/o British Biotechnology Products Ltd., 4–10 The Quadrant, Barton Lane, Abingdon, OX14 3YS, UK.

Kodak: Eastman Fine Chemicals 343 State Street, Rochester, NY, USA.

Life Technologies Inc., 8451 Helgerman Court, Gaithersburg, MN 20877, USA.

Merck

Merck Industries Inc., 5 Skyline Drive, Nawthorne, NY 10532, USA.

Merck, Frankfurter Strasse, 250, Postfach 4119, D-64293, Germany.

Millipore

Millipore (UK) Ltd., The Boulevard, Blackmoor Lane, Watford, Herts WD1 8YW, UK.

Millipore Corp./Biosearch, P.O. Box 255, 80 Ashby Road, Bedford, MA 01730, USA.

Molecular Probes Europe BV, PoortGebouw, Rijnsburgerweg 10, 2333 AA Leiden, The Netherlands

Molecular Probes, Inc., 4849 Pitchford Avenue, Eugene, Oregon 97402-9144, USA.

Neosystem Laboratoire, 7 Rue de Boulogne, 67100, Strasbourg, France.

New England Biolabs (NBL)

New England Biolabs (NBL), 32 Tozer Road, Beverley, MA 01915–5510, USA.

New England Biolabs (NBL), c/o CP Labs Ltd., P.O. Box 22, Bishops Stortford, Herts CM23 3DH, UK.

Nexins Research B.V., P.O. Box 16, 4740 AA Hoeven, The Netherlands

Nikon Corporation, Fuji Building, 2–3 Marunouchi 3-chome, Chiyoda-ku, Tokyo, Japan.

Perkin-Elmer

Perkin-Elmer Ltd., Maxwell Road, Beaconsfield, Bucks. HP9 1QA, UK.

Perkin-Elmer Ltd., Post Office Lane, Beaconsfield, Bucks, HP9 1QA, UK.

Perkin-Elmer-Cetus (The Perkin-Elmer Corporation), 761 Main Avenue, Norwalk, CT 0689, USA.

Pharmacia Biotech Europe Procordia EuroCentre, Rue de la Fuse-e 62, B-1130 Brussels, Belgium.

Pharmacia Biosystems

Pharmacia Biosystems Ltd. (Biotechnology Division), Davy Avenue, Knowlhill, Milton Keynes MK5 8PH, UK.

Pharmacia LKB Biotechnology AB, Björngatan 30, S-75182 Uppsala, Sweden.

Promega

Promega Ltd., Delta House, Enterprise Road, Chilworth Research Centre, Southampton, UK.

Promega Corporation, 2800 Woods Hollow Road, Madison, WI 53711–5399, USA.

Qiagen

Qiagen Inc., c/o Hybaid, 111–113 Waldegrave Road, Teddington, Middlesex, TW11 8LL, UK.

Qiagen Inc., 9259 Eton Avenue, Chatsworth, CA 91311, USA.

Schleicher and Schuell

Schleicher and Schuell Inc., Keene, NH 03431A, USA.

Schleicher and Schuell Inc., D-3354 Dassel, Germany. Schleicher and Schuell Inc., c/o Andermann and Company Ltd.

Shandon Scientific Ltd., Chadwick Road, Astmoor, Runcorn, Cheshire WA7 1PR, UK.

Sigma Chemical Company

Sigma Chemical Company (UK), Fancy Road, Poole, Dorset BH17 7NH, UK.

Sigma Chemical Company, 3050 Spruce Street, P.O. Box 14508, St. Louis, MO 63178–9916.

SLM-AMINCO, 810 W. Anthony Drive, Urbana, IL 61801, USA.

Sorvall DuPont Company, Biotechnology Division, P.O. Box 80022, Wilmington, DE 19880–0022, USA.

Stratagene

Stratagene Ltd., Unit 140, Cambridge Innovation Centre, Milton Road, Cambridge CB4 4FG, UK.

Strategene Inc., 11011 North Torrey Pines Road, La Jolla, CA 92037, USA.

United States Biochemical, P.O. Box 22400, Cleveland, OH 44122, USA.

Wellcome Reagents, Langley Court, Beckenham, Kent BR3 3BS, UK.

Index

Index